Regional Environmental Management

REGIONAL ENVIRONMENTAL MANAGEMENT

Selected Proceedings of the National Conference

Edited by

L. EDWIN COATE
PATRICIA A. BONNER

Conference Sponsors

San Diego County Integrated Regional Environmental Management Project and United States International University Center for Urban and Human Development

A WILEY-INTERSCIENCE PUBLICATION

JOHN WILEY & SONS, New York • London • Sydney • Toronto

Library of Congress Cataloging in Publication Data:

National Conference on Regional Environmental Management, San Diego, Calif., 1973.
Regional environmental management.

"A Wiley-Interscience publication."
Includes bibliographical references and index.
1. Environmental policy—United States—Congresses.
2. Regional planning—United States—Congresses.
I. Coate, L. Edwin, 1936– ed. II. Bonner, Patricia A., 1944– ed. III. San Diego Co., Calif. Integrated Regional Environmental Management Project. IV. United States International University. Center for Urban and Human Development. V. Title.

HC110.E5N35 1973 301.31'0973 74-26947
ISBN 0-471-16138-1

Printed in the United States of America

10 9 8 7 6 5 4 3 2 1

Conference Organization

Chairman: **Dr. L. Edwin Coate,** San Diego County Integrated Regional Management Project

Co-Chairman: **Dr. Thomas A. MacCalla,** United States International University

Conference Manager: **Patricia A. Bonner,** IREM Project

Session Chairmen:

Validity of the Regional Concept: **Dr. William Felling,** Ford Foundation

Regional Planning and Environmental Management: **John Tooker,** California Office of Planning and Research

Environmental Impact Statements: **William Matuszeski,** President's Council on Environmental Quality, and **Nicholas Yost,** Office of the Attorney General, California

Environmental Quality Indicators: **Terry Davies,** President's Council on Environmental Quality

Economic Analysis: **Dr. Ralph d'Arge,** University of California at Riverside

Information Systems and Modelling: **Dr. Peter House,** United States Environmental Protection Agency

Role of Special Interest Groups and Industry: **James Dale,** Industry Environmental Council

Environmental Education: **Dr. Thomas A. MacCalla,** United States International University, Center for Human and Urban Development

Preface

On February 26–27, 1973, in San Diego, California, two hundred men and women from across the country met to discuss regional environmental management—a topic that only a handful of people recognized before 1970.

The conference sponsors, San Diego County's Integrated Regional Environmental Management (IREM) Project and the United States International University's Center for Urban and Human Development, believed that it was time to bring the nation's regional environmental management specialists together. The issues appeared to be sufficiently well articulated to permit a conference on the technical and management aspects of the regional approach to be successfully conducted with significant benefits to the national environmental management community.

The emphasis of the conference in San Diego was practical, not philosophical. Presenters were asked to develop material that could be particularly useful in policy making and problem solving at the regional level. Material was to assist in these major functions: (1) determining the validity of the regional (both physical and political) concept of environmental management in terms of its function, scope, and relationship to higher and lower governmental entities; (2) identifying the principles of operation, organization, and intergovernmental relationships that are productive; and (3) identifying problem-solving techniques, system design concepts, laws, and other environmental management techniques and tools of particular relevance at the regional level.

Overall the conference had these objectives:

- To provide a forum for the exchange of information and working concepts.

- To enable those practicing environmental management to meet and discuss significant study results in both technical and management areas particularly useful at the regional level.
- To bring together publishable material, thus advancing the state of the art in the emerging field of environmental management.

To achieve these purposes, eight sessions were organized:

1. Validity of the Regional Concept
2. Role of Economic Analysis in Regional Environmental Management.
3. Environmental Quality Indices.
4. Information Systems and Modeling.
5. Environmental Education.
6. Role of Special Interest Groups and Industry in Regional Environmental Management.
7. Environmental Impact Statements.
8. Regional Planning and the Regional Environmental Management Process.

The sessions were not meant for "specialists" in the areas discussed. Other conferences cover economics for economists, modeling for modelers, and planning for planners. This conference was for people relatively new to regional environmental management, the managers and practitioners who wanted to know more about the total operations of the field and how the many diverse disciplines can be integrated and interrelated.

These Proceedings are selections from the National Conference on Regional Environmental Management presentations and other submitted papers. The book is a beginning point for readers like the attendees—people who want to know how to begin to plan and bring about a better quality of life.

PATRICIA A. BONNER
L. EDWIN COATE

Windsor, Ontario, Canada
Seattle, Washington
September 1974

Contents

Regional Environmental Management

Part 1

Validity of the Regional Concept

Patricia A. Bonner

Ms. Bonner, Managing Editor of these Proceedings, is the Information Officer of the Great Lakes Regional Office of the International Joint Commission, United States and Canada. Formerly she was IREM's Communications Director.

Today's environmental problems are a result of yesterday's piecemeal management of an intricately woven fabric of air, water, land, and the inhabitants of these regions. Management of environmental resources appears to be a highly defined and functionally systematic governmental activity. Indirectly, environmental management has been occurring for some time, but in a highly diffused manner, lacking a systematic basis or set of guideline criteria.

The environment is everywhere and cannot logically be carved up by jurisdictional boundaries or managed by single-purpose governmental authorities such as public health, highways, flood control, water quality control, or solid waste management. In the past, responsibility for care of the environment was diffused across a myriad of authorities; each one attempted to grapple with that portion of the environment within its jurisdiction, and each structured efforts through guidelines tailored by lines on a map and single-purpose goals, perceptions, and professional beliefs. Agency activities at all governmental levels have both a direct and an indirect effect on environmental resources, and in all probability this trend will continue for many years.

The landscape of environmental management is filled with complex interrelationships not only between natural entities and evolutionary processes, but also between human activities, governmental concerns, and resource management programs. Potentially, every agency has some effect on the environment.

To effectively perform management of the environment, there must be integration of the single-purpose groups: their outlooks must cross geographic and jurisdictional lines, permitting interrelated activities to be performed jointly, cooperatively, in parallel, or in correct sequence, rather than in competition, out of phase, or in a communications vacuum. Similar spheres of activity (e.g., water districts, a river system, cities, school districts, and planning units) should relate to one another. Decisions having an impact on huge geographic areas should be made jointly by all affected jurisdictions. These are arguments for "regional" control. But what is a region?

An air basin, one city, a city crossing state or even international boundaries, a county, a few counties, a state, several states, are all "regions." Can they operate effectively to perform environmental management?

Regionalism in EPA: From Programs to Policy

GARY H. BAISE

It is appropriate for us to make the United States bicentennial an occasion to reexamine our institutions and, where appropriate, to oil the machinery and see that it stays in good repair. And that is why we are currently devoting so much attention to our federal system.

Federalism

A generation ago, many sophisticated people thought that federalism was dead and that the states were anachronisms. We have tended to take the federal system for granted. It is only when good things are threatened that we begin to measure their true worth, to see how indispensable they are. We are at last beginning to appreciate the true worth of the federal system—probably because it was in jeopardy for so long.

Our federal system is our contribution to the world's political theory. It is uniquely American; a reflection of the fact that we like our power plural.

The friendly and not-so-friendly rivalries of federal, state, and local governments restrain absolute power—a desirable end in itself. But more than that, the rivalry with all its internal clashes and competition creates a dynamic tension that in the end promotes good government.

For years, because of national emergencies like the great depression and the world wars, the states were less powerful than the national government. Now we are trying to restore the balance, to achieve equipoise—that is really the

Mr. Baise was assistant to the Administrator, Director of Office of Legislation, Environmental Protection Agency (EPA). He is now in private law practice.

goal of the "new federalism" and revenue sharing, the federal regional councils, the grant consolidation program, and federal assistance review (FAR). Other examples of means to achieve the end are the whole government reorganization plan, and EPA's own reorganization policy.

I think one of the happiest developments of our epoch is that republicans and democrats, liberals and conservatives (by admittedly different routes and as the end product of different historical processes) have arrived at a consensus that affirms that the federal government is not—nor should it seek to be—all things to all men.

President Nixon said in his second inaugural address in January 1973:

> We have lived too long with the consequences of attempting to gather all power and responsibility in Washington. Abroad and at home, the time has come to turn away from the condescending policies of paternalism—of "Washington knows best."

For years, we tried to visit all our troubles and problems on Washington, and we found very spotty performance in many cases. We found all too often that if you want something done, Washington was the last place to go and that in many cases the state, local, and regional governments represent the appropriate levels for action—that is, the levels that can be more responsive to the people's needs and more attentive to their requirements.

That is also the point of the president's plans to reorganize the federal government. The president intends to provide the people with good organizational structures that will allow and encourage them to get on with the jobs at hand.

That is the point of the Federal Regional Council, which brings together the federal agencies in the 10 federal regions to meet regularly and share their problems.

In the process, we hope to cut down duplication, to spike contradictory programs before they get off the ground, and to consolidate and systematize the approach of the federal government in its relations with state and local governments.

New Federalism and EPA

I am very proud of the way EPA has structured itself to complement the president's policies under the "new federalism." We have consciously and deliberately tried to reverse the flow of power to Washington and to bring decision making closer to the people. We were lucky in many ways because ours is a new agency and we were able to shape its future. Other, older agencies find such planning infinitely more difficult. In a very important way, the decentralization of EPA is an extension of the president's policy that led to the creation of the agency. In July 1970 when he submitted Reorganization Plan 3 to the congress, the president affirmed that "the Executive Branch of government should be organized around basic goals."

Our mission—to protect and enhance the environment—has also placed us in the vanguard of this movement. The statutes under which we function almost universally require a close and continuing relationship and interdependency between federal, state, and local governmental units. We accept that responsibility and rejoice in it. (Former) EPA Administrator William Ruckelshaus ordered that the agencywide intergovernmental relations functions become a part of my own office of legislation, and we are working to open lines of communications to the governors, state legislators, mayors, and municipal officials, directly and through their representatives in Washington, the National Governors' Conference, the Council of State Governments, the National Legislative Council, and similar groups. We are cranking those elected representatives of the people and their organizational representatives into our regulation drafting process at the beginning, and the response has been gratifying. We intend to keep on doing this and to make what heretofore has been the exception into the rule of the future.

We have physically decentralized: more than 70 percent of EPA's employees are in the field, 3100 in the regional offices and 3400 in National Environmental Research Centers, in laboratories, and other field locations. In each of the 10 federal regions, EPA is headed by a regional administrator who reports directly to the administrator. But this is not a paper situation. We mean what we say and there has also been a meaningful decentralization of authority.

Our regional administrators have the authority to act. Washington establishes the basic policy and procedural guidelines; the regional administrators carry out policy, and there is no second guessing. We reason that the regions are best suited to respond with sensitivity and imagination to the variables and subtleties of state and local priorities. It remains for Washington to evaluate the work in the region and to measure performance.

The president charged EPA to take a holistic approach to the environment, to view it not as isolated series of events or pollution situations but rather, as an entity in which disciplines overlap and all segments are interrelated and interwoven. This "systems approach" or "policy approach" rejects as simplistic the notion that we can protect the environment or enhance the quality of life by stacking a series of separate airtight Washington-based programs on top of one another.

Let me give you an example. The president in his environmental message of February 15, 1973 made a national land use policy bill a cornerstone for his program in this congress and put its passage high on his priority list. In doing so, he stipulated that land use policy is a basic responsibility of state and local governments, and it is understood that the states and localities must make the basic decisions affecting their land. The president's proposal would encourage the states, in cooperation with local units, to legislate a rational and coherent land use policy, one that would protect lands of critical environmental concern and regulate the siting of important new facilities such as airports and highways, as well as major private developments.

In recent years, we have seen the effects of runaway growth, growth that is unchanneled and, some critics claim, incoherent. Developments and tracts have been built with no attention to the needs of the residents—schools for their children, transportation, shopping facilities. Those essential services and social needs cannot be overlooked when there is a systematic and coherent land use policy in effect and when planning preceeds action.

Planning is the key to the president's land use proposal—not, the rigid, insensitive, procrustean planning as it is known in the socialistic world. Rather, we mean planning in the highest and best sense: To make conscious and rational decisions about how and when we want land, our precious resource, used and developed.

Under the administration's proposal, there will be certain funds available to the states for the planning itself, but the bill has a sticker: if the states opt to have no land use program at all, they stand to lose from 7 to 21 percent of future federal highway trust funds, airport development funds, and water conservation funds. It is a safe bet that if this bill becomes law, the states will promptly enact land use policy legislation—it will be in their interest to do so.

The bill is not, I stress, a zero growth proposition. Growth we must and shall have, but let it be growth that is rational and not chaotic.

I am confident that a national land use policy bill will be passed, eventually, once the state statutes are enacted, the federal government can bow gracefully out of the picture. That is the kind of creative federalism the president favors and the country needs: not programs in which every jot and tittle is decreed by a nameless regulator in Washington, but a prod to policy, to move the states and local and regional groupings to accept the responsibility to make decisions.

That responsibility can be tough, and a lot of flack goes with it as you well know, and a lot of hard decisions and headaches. But it is worth the effort.

Editors' note: The Land Use Policy Act encountered great opposition in 1973, and in 1974, was again defeated.

An Analysis of Institutional Forms

L. EDWIN COATE

In response to widespread public concern over the increasing deterioration of environmental quality, as well as the perceived need for governmental action to alleviate the situation, significant changes took place in environmental management structures at various governmental levels beginning early in 1970.

The federal government created the Environmental Protection Agency to consolidate more than 15 environmental programs that had been scattered among different agencies. Early in 1970 the President's Council on Environmental Quality was established as an advisory body to coordinate federal policy and action in the environmental field.

Many states in 1970 reorganized to create their own type of "Department of Environmental Protection." Illinois set up three wholly new agencies, each responsible for a different facet of environmental management. The State of Washington's Department of Ecology combined four previous departments (Water Pollution Board, Air Pollution Board, Water Resources, and Solid Waste Management) into a new, functionally integrated environmental "superdepartment."

On the regional level there were Councils of Governments, representing a somewhat new form of institutionalized cooperation in environmentally related activities on a regional basis; however, the councils had strictly advisory functions. Regional efforts, such as the Bay Conservation and Development Commission, the Metropolitan Council of the Twin Cities, and the Tahoe Regional Planning Commission, are recent innovative regional efforts to implement en-

Dr. Coate, former IREM Project Director, is now Deputy Administrator for Region X, Environmental Protection Agency.

vironmental programs. They have recognized the importance of relating land use and transportation planing to pollution regulation.

Because of lack of funds, staff, or time, local governments have made few attempts to reorganize. In this event, environmental management cannot occur effectively at the local government level because environmental management problems are regional. Problems of air and water pollution, solid waste management, and land use planning require multijurisdictional solutions.

IREM: A Case Study in San Diego County

During the 1960s the Ford Foundation showed increasing interest in funding environmentally related projects. The Foundation had traditionally invested in environmental research studies, but in 1970 it decided to try a new approach. It made a grant to a government agency involved in environmental activities on a regional basis. In April 1971, the Ford Foundation awarded the County of San Diego a $725,000 grant for the two-year project on Integrated Regional Environmental Management (IREM).

To understand the IREM experience, some background information on San Diego County is necessary. Most identifiable environmental problems within the county are closely related to the overall phenomenon of population growth. The fundamental problem is directing the pattern of land use, particularly of new housing needed to accommodate projected population. Economic interests are now pressing to add to the rate of growth through the greatly increased submission of development plans.

The principal environmental problem identified by the people of the region is air pollution. Meteorological and topographical factors in the San Diego region make the area highly susceptible to problems associated with air pollution. In a five-year study, San Diego had *60 percent more days* than Los Angeles when weather conditions could have led to air pollution episodes.

The county's chief consumable natural resource is its undeveloped land. Major parts of that resource are likely to be irreversibly committed to a long-term land use pattern without consideration of environmental effects beyond the boundaries of individual development projects. In the near future, many major land use decisions for San Diego will have to be made.

The overall purpose of the IREM project was to determine what environmental management was and how effectively it could be accomplished on a regional level from within a local government. This included determining appropriate programs and regional institutional relationships. The project's specific purposes were to work within the county government framework to:

1. Respond to federal and state legislative requirements.
2. Mobilize community resources and support decision makers.

3. Respond to citizen concerns as articulated by the Board of Supervisors.
4. Enhance the regional image.
5. Provide a rational approach to environmental issues.

In light of these general and specific purposes, the IREM project concluded that the following set of goals and objectives were most appropriate:

1. The Identification of Natural Resources.
2. The Integration of Resources for Action.
3. Provision of Information to Decision Makers.
4. Development of New Technologies.
5. Development of Management Processes and Programs with National Application.
6. Validation of Environmental Standards.
7. Development of a Regional Project Management Capability.
8. Identification of Needs for Institutional Change.

The internal organizational structure of IREM evolved to allow close examination of the types of organizational attributes that best supported the goals of a regional environmental management organization.

The final organization plan used by IREM divided the project into two functional areas: (1) A Program Management activity with responsibilities for policy development and for the technical management of research, development, and demonstration projects; and (2) a Regional Environmental Services activity to provide a capability for reacting to immediate issues, for implementing pilot programs, and for community service (see Figure 1).

The Program Management activity was further divided into sections dealing with land use analysis, pollution management, and communication and evaluation. The Regional Environmental Services activity contained sections dealing with economic analysis, community involvement, and environmental impact analysis. Table 1 illustrates the relationships between the final organizational structure and the 23 program areas identified at the project's outset.

IREM conducted its activities using its own staff (approximately 30 persons), the staff of other governmental agencies under interagency agreements, staff consultants, and contracts with universities and consulting firms.

The project was located in the Environmental Development Agency, the land use and transportation planning agency of San Diego County. This agency, created in early 1971 during a county reorganization, consolidated various environmentally related functions, including both regional planning and zoning. The Environmental Development Agency was located under the county's Administrative Officer, who is directly responsible to the Board of Supervisors. Because the project was somewhat isolated from other public and

Table 1. Relation Between the Organization of the Project and Programs

Programs	Program Management Division	Communication and Evaluation Section	Land Use Analysis Section	Pollution Management Section	Regional Services Division	Environmental Impact Analysis Section	Community Involvement Section	Economic Analysis Section
Natural resources inventory			•					
Monitor environment				•				
Early warning			•	•		•		
State-of-the-environment indices		•					•	
Land law review			•					
Ecological zoning			•			•		
Coastline development policy			•					•
Lagoon development policy			•					•

Air pollution				●		●		●
Water pollution				●		●		●
Noise pollution				●		●		●
Pesticides				●		●		●
Solid waste disposal				●		●		●
Energy				●		●		●
Visual aesthetics			●					●
Economic modeling and evaluation			●			●		●
Environmental Impact Statement process					●	●		
Computer graphics			●					
Manpower training	●				●			
Public information		●					●	
Public education		●					●	
Proposals for funding	●						●	
Community action							●	

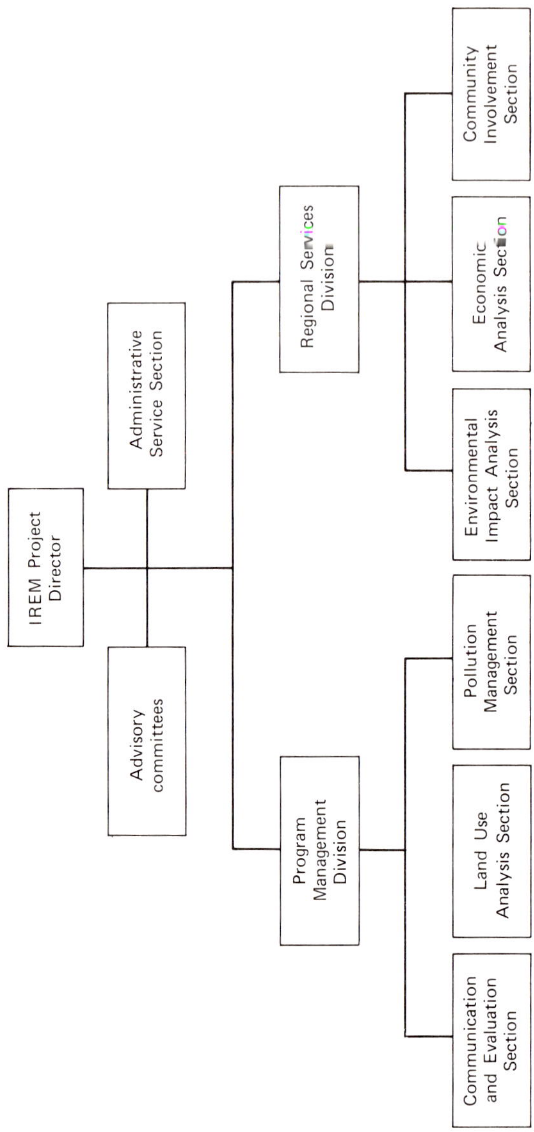
IREM Project Director
Advisory committees
Administrative Service Section
Program Management Division
Regional Services Division
Communication and Evaluation Section
Land Use Analysis Section
Pollution Management Section
Environmental Impact Analysis Section
Economic Analysis Section
Community Involvement Section

Figure 1

private environmental institutions in the San Diego region, it soon became evident that achieving a true *regional* environmental management capability would require the establishment of definite linkages between IREM and other environmental institutions in the region. To develop a strategy for building these linkages, the environmental issues and problems facing the region were identified. Once IREM had determined the agencies to which it should become formally linked, program development began. These programs provided the series of transitions through which IREM would slowly and incrementally build the necessary institutional bonds within the region.

During the early months of the project, IREM identified 23 major program study areas. Of these, some were quickly discarded as inappropriate or infeasible for IREM; others were developed to various degrees of complexity. One major effort was a regional Natural Resource Inventory to document natural resource information essential for land use and public works decision making, as well as for long-term planning, environmental impact review, and measurement of our current regional environmental status. Computer graphics techniques developed by IREM were put to use on several regional land use projects. IREM staff prepared a series of issue papers on environmental problems facing the region. Grant applications to attract additional funds to assist in solving complex problems of regional environmental significance were also prepared.

At the end of its first year, IREM staff had concentrated its program areas down to nine, all selected for more intensive work because they had economic, environmental, and social implications for regional environmental management. The programs selected include the following:

1. Environmental Impact Statement Review.
2. The San Diego Clean Air Project, funded by a $425,000 grant from the Environmental Protection Agency, focusing on alternative courses by which the county could meet the 1975 air quality standards.
3. The development of an Environmental Element for the General Plan, beginning with coastline management and open space data.
4. State-of-the-environment reporting, to produce a status report on the environmental condition of the region, through a set of environmental quality and indicators.
5. A joint city/county economic analysis of regional growth, aimed at determining the economic effects of new large-scale subdivisions on the region, and representing the establishment of an institutional link between IREM and the City of San Diego.
6. University collaboration—under a $130,000 grant from the National Science Foundation, a series of five research projects was coordinated, not only to provide solutions to regional environmental problems, but also to test the ef-

fectiveness of local government–university collaboration for such problem solving.

7. Public information, including a monthly environmental education newspaper for the county schools and natural resources publications.

8. Institutional analysis, involving a proposed reorganization of the county's Environmental Development Agency.

9. Issue analysis for decision makers especially responding to environmental issues at the request of the Board of Supervisors.

Through these programs, at the end of 18 months of its project life, IREM had developed a strong network of institutional linkages in the region, as well as with the state and federal governments (see Figure 2). Institutional relationships developed by the IREM Project can be characterized by the following:

1. Program collaboration.
2. Financial assistance.
3. Information exchange.
4. Issue analysis.

What was learned about regional environmental management from the IREM experience? The IREM case study allows us to conclude that to be most effective, regional environmental management must relate agencies involved in land use and transportation planning to those which are responsible for regulatory pollution management functions. An effective regional environmental management agency must have linkages to all the key environmentally related institutions in the region.

The IREM experience also led to the conclusion that effective regional environmental management had to be carried out with some type of authority or under a definite mandate. Advisory functions and research and development functions were necessary and important but proved to be inadequate by themselves. IREM, located as it was in a county government, could not effectively fulfill all these criteria.

In light of the IREM experience and what has occurred at the other governmental levels, it is possible to design a model of a more effective institutional form for regional environmental management.

Regional Environmental Management: A Model

The model proposed focuses on a Regional Environmental Board, the jurisdiction of which is a *natural* region. This Regional Board would report to a State Environmental Board. The Regional Board would have advisory and regulatory functions and would engage in research and development. This model

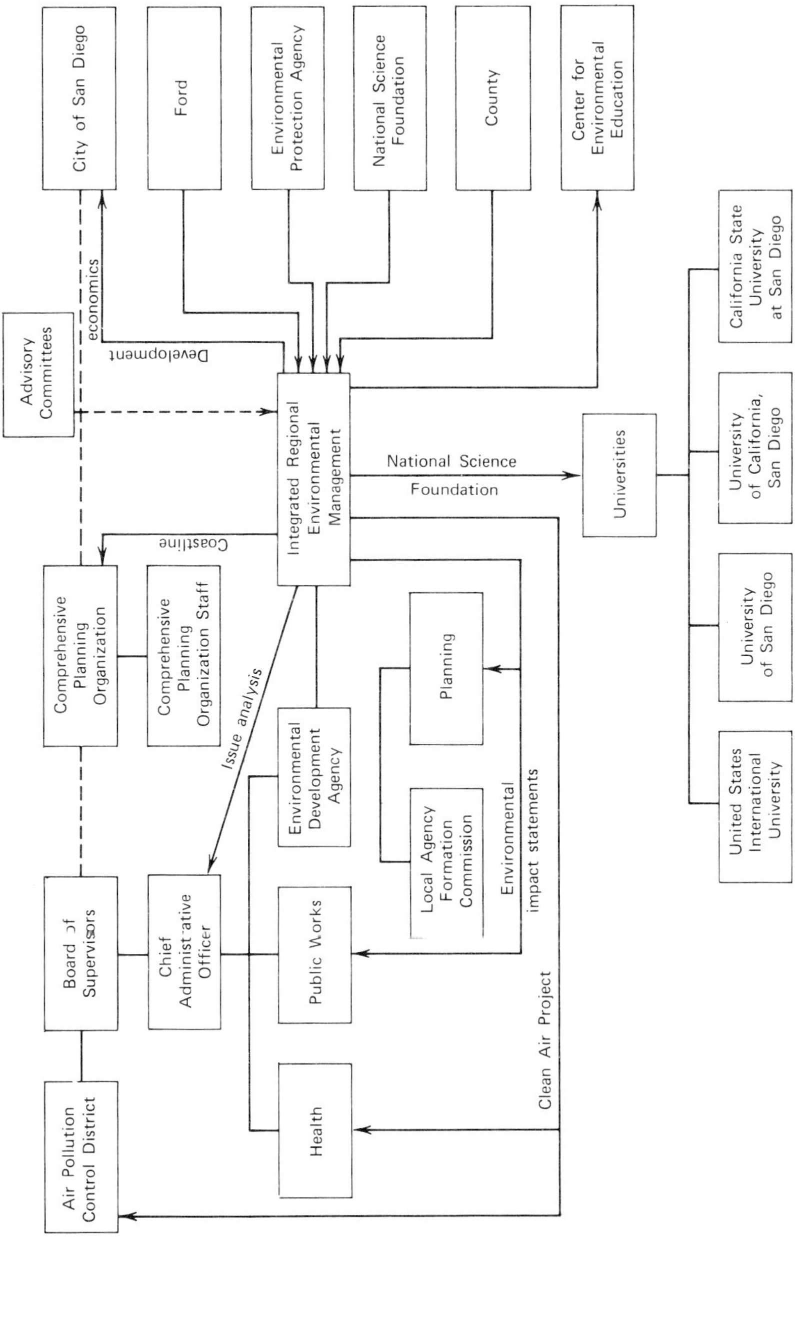
City of San Diego
Ford
Environmental Protection Agency
National Science Foundation
County
Center for Environmental Education
Advisory Committees
Development economics
Integrated Regional Environmental Management
National Science Foundation
Universities
California State University at San Diego
University of California, San Diego
University of San Diego
United States International University
Coastline
Comprehensive Planning Organization
Comprehensive Planning Organization Staff
Issue analysis
Environmental Development Agency
Planning
Local Agency Formation Commission
Environmental impact statements
Board of Supervisors
Chief Administrative Officer
Public Works
Clean Air Project
Air Pollution Control District
Health

Figure 2

would provide for a consolidation of the planning and management functions of land use and transportation agencies with those of agencies responsible for pollution regulation.

The Regional Board would be divided into a series of departments whose functions and structures would provide for the aspects of regional environmental management identified in the conclusions listed previously. The structure of this model appears in Figure 3.

The Regional Environmental Board, the key element of the proposed model, would be composed of six members: three appointed by the State Board and three appointed by a Regional Selection Committee. The Selection Committee would include two members from any county Board of Supervisors having jurisdiction over part of the area, the mayor of any city lying partly or wholly in the region, and three Special District Representatives selected as a result of a conference of Special Districts in the region. The Selection Committee would choose three persons to represent their region as a whole.

The six Regional Board members would serve 6-year terms (staggered—one year each). One Regional Board member would have extensive experience in the field of pollution regulation and control; one would be an expert in conservation and natural resources management; and one would be a specialist in land use and transportation planning or environmental management.

Members of the Regional Board would be full-time appointees, meeting whenever necessary. The Regional Board would have the power to:

1. Conduct hearings and investigations anywhere in the region.
2. Recommend to the state legislature specific environmental proposals.
3. Carry out environmental management duties such as: (*a*) setting standards, (*b*) preparing plans, (*c*) issuing cease and desist orders, (*d*) denying building permits in environmentally sensitive areas, (*e*) requiring local agencies to assist in regional environmental management activities, and (*f*) reviewing regulations set forth by local agencies.
4. Review projects with regional significance, test them against established regional environmental policies and objectives and, if appropriate, disapprove the projects or require their modification.
5. Review and disburse federal, state and local funds.
6. Contract for services and appoint groups.
7. Enforce powers by fining individuals and groups when necessary.

The Regional Board would prepare an annual state-of-the-environment report detailing the status and conditions of environmental problem areas, including a discussion of programs under way to improve the environment in the region.

Although the Regional Board's departments can be separately and functionally described, the key to the success of the environmental management

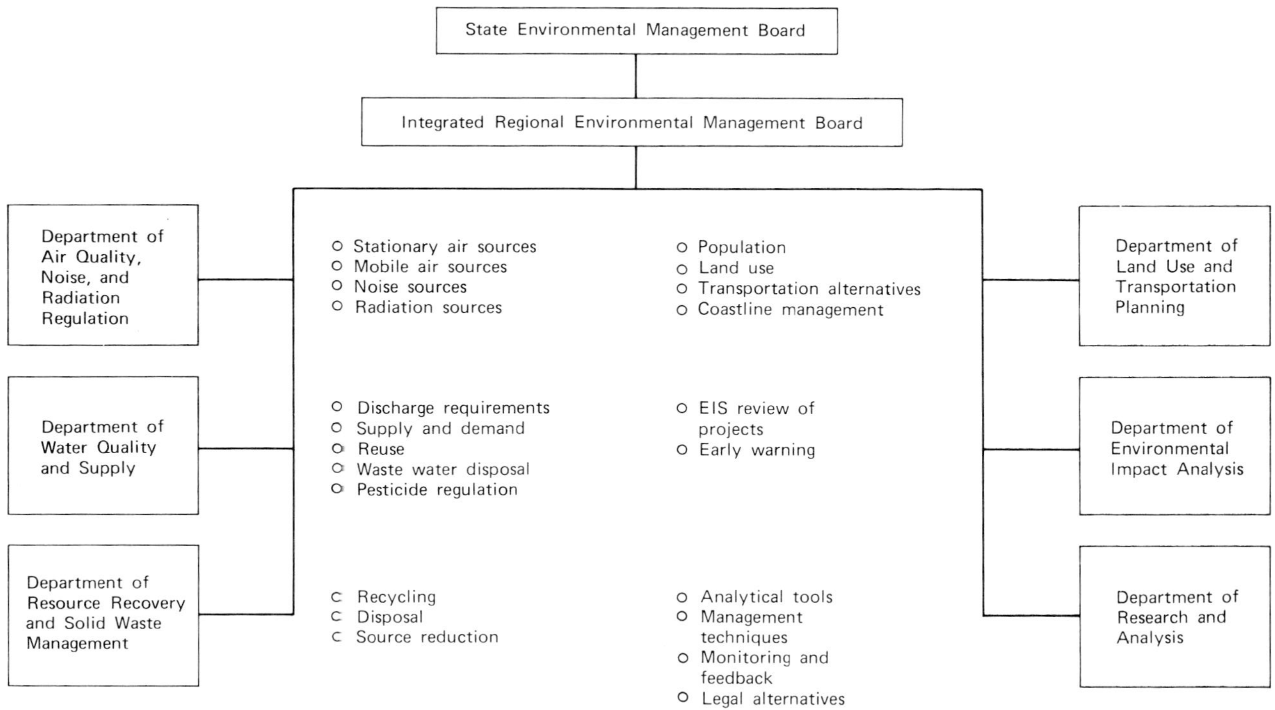
State Environmental Management Board
Integrated Regional Environmental Management Board
Department of Air Quality, Noise, and Radiation Regulation
Department of Water Quality and Supply
Department of Resource Recovery and Solid Waste Management
Stationary air sources
Mobile air sources
Noise sources
Radiation sources
Discharge requirements
Supply and demand
Reuse
Waste water disposal
Pesticide regulation
Recycling
Disposal
Source reduction
Population
Land use
Transportation alternatives
Coastline management
EIS review of projects
Early warning
Analytical tools
Management techniques
Monitoring and feedback
Legal alternatives
Department of Land Use and Transportation Planning
Department of Environmental Impact Analysis
Department of Research and Analysis

Figure 3

model is the way the different departments integrate and relate their planning and management functions on a day-to-day basis.

The departments are as follows:

1. Air Quality, Noise and Radiation Regulation.
2. Water Quality and Supply.
3. Resource Recovery and Solid Waste Management.
4. Land Use and Transportation Planning.
5. Environmental Impact Analysis.
6. Research and Analysis.

The functions and programs of the various departments would undoubtedly vary according to local/regional circumstances. An example of how the model would look if implemented in the San Diego region is presented in Figure 4. The San Diego IREM case study showed the necessary institutional linkages, but it also revealed how hard it is to achieve such implemental and integrated powers at the regional level from within a county government.

Some type of regional authority is essential to implement the proposed model. Accordingly, four implementation alternatives are proposed: (1) a newly created Regional Agency mandated by the state, (2) a city/county merger, (3) a strengthened Council of Governments (with taxing powers), and (4) a regional center for Environmental Coordination.

State-Mandated Regional Agency

Specific examples of the state-mandated regional agency approach already exist: Assembly Bill 681 introduced into the California Legislature by Assemblyman Edwin L. Z'Berg to create a pollution management agency, and Proposition 20, the citizens' ballot initiative that created the Coast Regional Commission. Thus we have attempts to impose the regional model by the state.

City/County Merger

The model could also be implemented by a merger of existing local agencies. In trying to eliminate the usual duplication of efforts and conflicting decisions, the two major local government entities could consolidate through the creation and adoption of a city/county charter. This type of merger has numerous well-known precedents such as Nashville–Davidson in Tennessee, Miami–Dade in Florida, Indianapolis, and San Francisco.

The regional government created would be able, by merging various departments and resources, to create a "superagency" that would have equivalent functions to the Regional Environmental Board proposed in the model.

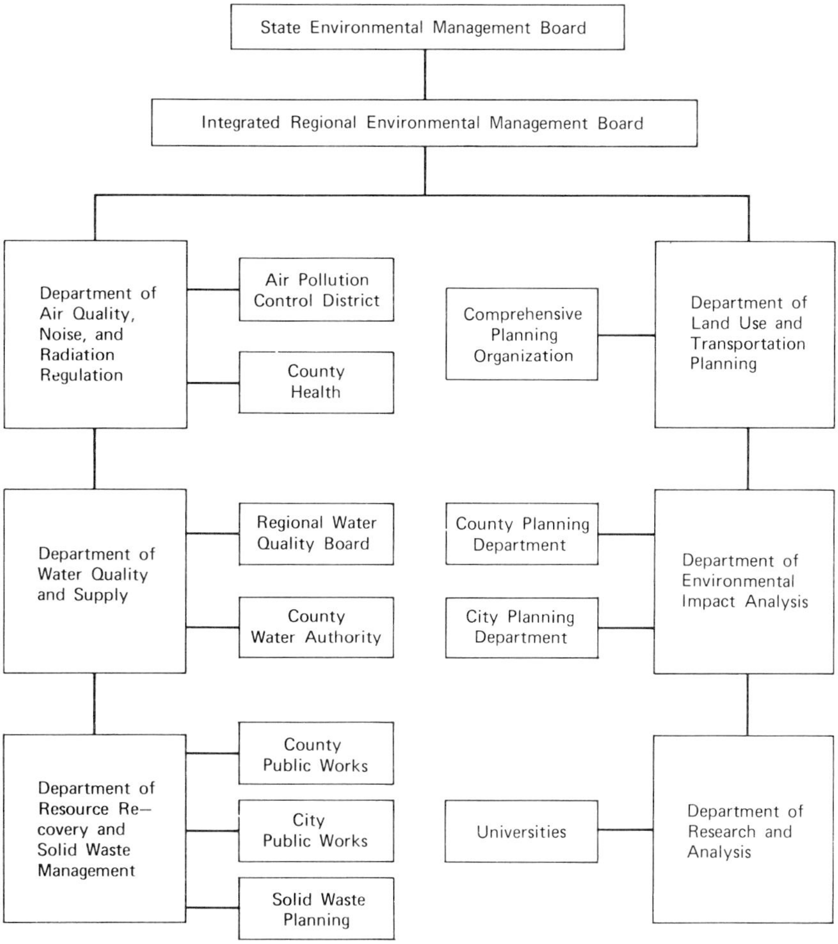

Figure 4

Strengthened Council of Governments

A stronger regional Council of Governments would simply acquire broad implementation authority over pollution management agencies, as well as implementation powers over the land use and transportation functions it already has. It would be able to review each local agency's capital improvements program to make sure plans were beneficial to the region as a whole. It would also have taxing powers to make it independent of existing agencies in the region. An example of such an arrangement was Assembly Bill 3050, the San

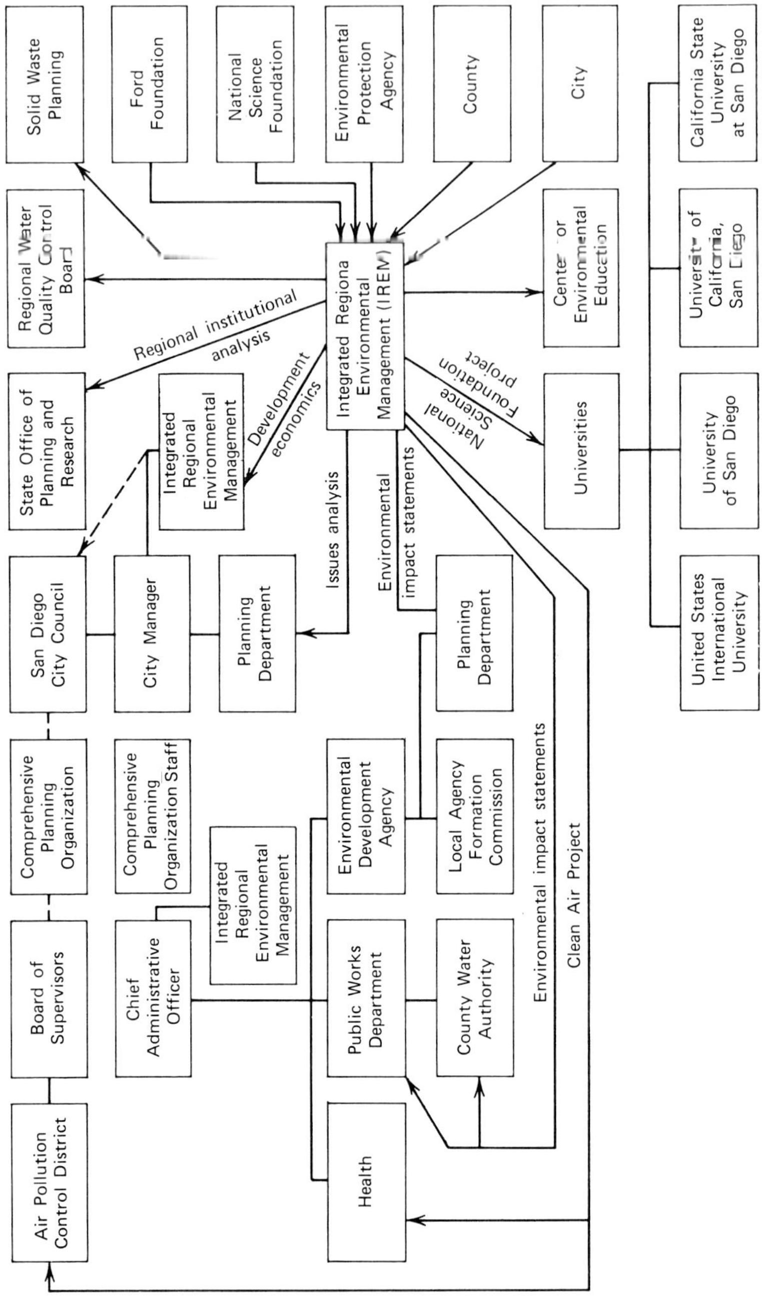

Solid Waste Planning
Ford Foundation
National Science Foundation
Environmental Protection Agency
County
City
California State University at San Diego
Regional Water Quality Control Board
Integrated Regional Environmental Management (IREM)
Center for Environmental Education
University of California, San Diego
Regional institutional analysis
State Office of Planning and Research
Integrated Regional Environmental Management
Development economics
National Science Foundation project
Universities
University of San Diego
Issues analysis
Environmental impact statements
San Diego City Council
City Manager
Planning Department
Planning Department
United States International University
Comprehensive Planning Organization
Comprehensive Planning Organization Staff
Integrated Regional Environmental Management
Environmental Development Agency
Local Agency Formation Commission
Environmental impact statements
Clean Air Project
Board of Supervisors
Chief Administrative Officer
Public Works Department
County Water Authority
Air Pollution Control District
Health

Figure 5

Diego Council of Governments Act, introduced into the 1971 California Legislature by then-Assemblyman Pete Wilson, now Mayor of the city.

Regional Center for Environmental Coordination

A bureaucratic solution if none of the foregoing alternatives is immediately possible is the regional center. This interim phase model would be a logical modification of the IREM project and its institutional relationships until a true regional model could be implemented.

The principal concept is the establishment of a regional center for environmental coordination, which would consist of a captive, nonprofit research institute with counterpart offices located in the major governments in the region. A link to the state government is also critically important (see Figure 5).

The regional center would have a multiplicity of funding sources, including federal (e.g., Environmental Protection Agency, National Science Foundation), state, and local funding, or at least office space and support. The center would provide environmental issue analysis for the local governments involved.

It is assumed that the center would carry out, from a neutral research standpoint, programs similar to those found to be appropriate in the IREM experience. It would also attempt to coordinate the planning and management functions of the various environmentally related agencies in the region.

Conclusion

The regional concept for environmental management is not only valid, it is essential. We have finally begun to question the efficacy of institutions that deal with environmental quality at all governmental levels. The resulting analysis leads to the conclusion that a new type of regional institution must be created. The IREM experience is a beginning that yielded significant information about such a regional institutional form. The functions and attributes finally described in the hypothetical model may not be absolutely accurate, but they do begin to approximate a feasible structure for actualizing regional environmental management. It is now up to the practitioners—those involved in local government—to recognize that such new regional organizations are possible and must be brought into being if a rational compatibility is to be found between man and his natural environment.

Some Thoughts About Regions and People

JOSEPH E. BODOVITZ

Regional Identity

Fifteen years ago, in 1958, a university researcher from Berkeley was in Venice (Italy, not Los Angeles) completing a major book about a major region. The book, *The San Francisco Bay Area: A Metropolis in Perspective*, was written by Mel Scott.

> To be great in the same way that Venice is Great [Scott states], the San Francisco Bay Area must now and in the future have residents who identify themselves with their metropolitan regional environment as closely as generations of Venetians have with theirs. . . . If American metropolitan communities are to become worthy of the love of their citizens, something of the unifying purpose that inspired the Venetians must be applied in their development . . . there must be a vision of magnificent possibilities.
>
> Vision comes from decades of shared experiences; and in metropolitan regions which are continually adding new residents to their populations, the development of common goals becomes a problem in itself. . . . Swift augmentation of the means of attaining goals causes relatively rapid shifts in the goals themselves, (and) consensus becomes all the more difficult. Should we conclude, then, that it is well-nigh impossible for the people of the metropolitan region to agree on what that region can become and to work toward a common goal? To accept such a conclusion would be to deny the capacity of Americans in general and Californians in particular for dealing with the formidable but invigorating challenges of their times.

I believe Scott's statement is important for three reasons. First, because, perhaps ironically, the smog, water pollution, and other environmental prob-

Mr. Bodovitz was named Director of the California Costal Commission in 1973. He had been Director of the San Francisco Bay Conservation and Development Commission.

lems that bring us together are affecting Venice and indeed almost every region in the industrialized world.

Second, because Scott correctly reminds us that although our current emphasis may be on the physical ills of a geographic region, we will not solve those problems if we think of a region as only a chunk of territory. If the natural resources of a region are to be well managed, it will be not only because planners and other technicans have done their jobs well, but also because the people of the region have somehow come to have a common appreciation of what has gone before in the region, of what might be in the future, and of what needs doing today.

This aspect of regionalism is more commonly recognized in Europe, with its longer history and older traditions. The 1960 report of the Royal Commission on Local Government in Greater London notes the controversy over protection of some of London's historic areas and concludes that the desire to preserve part of the past "is not mere sentimentality. The awareness of them shows that Londoners know something of what London means. To destroy or obscure these monuments of the past would be to undermine the foundations of the present."

In this country, or in California at least, it may be difficult or even impossible, to achieve the feeling of change within stability that exists, however uneasily, in Venice or London. Indeed, much of the environmental problem in California stems from the exploitation of land, air, and water by those who feel little or no stake whatever in the future of an area.

San Francisco's Response—B.C.D.C.

Sometimes it takes a serious threat to a major resource to mobilize the regional consensus necessary for accomplishment, and this brings me to my third reason for mentioning Mel Scott. Having completed his book in Venice, Scott returned to the Bay Area and published another book in 1963: *The Future of San Francisco Bay*. It described the slow, piecemeal destruction of the Bay by filling and noted the lack of controls on filling, as well as the need to treat the Bay as a single body of water that lies at the heart of a major metropolitan area. Scott believed that a Bay Conservation and Development Commission was called for. In 1965 the state legislature agreed and created such a commission, in a form slightly different from that suggested by Scott, but having the same responsibilities—to plan for the future of the Bay and to regulate both conservation and development affecting the Bay. It has often seemed to me that the most important word in the commission's name is the conjunction, not the nouns—it is the Bay Conservation *and* Development Commission, not the Bay Conservation *or* Development Commission.

Scott's book was one of the reasons for the creation of BCDC. The other principal factors were the unlikely combination of a powerful state senator, the wife of the president of the University of California, a disc jockey, and thousands of letter-writing, telegram-sending Californians.

The university president's wife, Mrs. Clark Kerr, was driving across the Bay Bridge from Oakland to San Francisco one day in the early 1960s with Mrs. Donald McLaughlin, whose husband was then chairman of the University's board of regents. The two women were discussing the just-announced plans of the City of Berkeley to enlarge itself by filling some 2000 acres of the Bay. "The Bay is certainly beautiful today," said Mrs. McLaughlin, looking at the miles of blue water. "Well," said Mrs. Kerr, "it won't even be a bay if every city does what Berkeley wants to do."

Both women were suddenly struck by the prospect of dozens of dredges, dumptrucks, and bulldozers at the water's edge, filling the shallow margins of the Bay with dirt and sand and debris, thus greatly shrinking the Bay. Could this really happen? Was the bay in danger? How much of it could be filled? Mrs. Kerr, Mrs. McLaughlin, and a third university wife, Mrs. Charles Gulick, decided to find out.

They learned to their surprise that almost one-third of the Bay as nature created it had already been diked off or filled in, and that more than half of the remaining bay was shallow enough to be economically fillable over the next half-century. ("Susceptible of reclamation" was how the Army Corps of Engineers had put it in a detailed study). And they learned that a rising population in the Bay Area would mean increasing pressures to fill more of the Bay for the same purposes it had been filled in the past—factories, homesites, airports, garbage dumps, roadways, and all the other needs of urban civilization.

Furthermore, nobody seemed to know the consequences of so much filling, and with 32 cities and 9 counties each having parts of the Bay within their boundaries, nobody was responsible for looking at the Bay as the single body of water it is.

The three women, unsure of what they could do as citizens in so complex a matter, recruited a few friends and formed the Save San Francisco Bay Association. They immediately began trying to interest governmental officials at all levels in doing something to protect the Bay. They badgered legislators in Sacramento to stop the piecemeal filling. And they began to find success: an Oakland assemblyman, Nicholas Petris, introduced a bill in 1964 that would have stopped the filling while a plan for the Bay was prepared. But Petris (who is now a state senator and still a leader in the fight for wise use of the Bay) found little support, and his bill was emasculated by the first committee to which it was sent.

Then the women sought the help of Senator J. Eugene McAteer of San

Francisco, a member of the powerful "old guard" in the state senate, a lawmaker with a reputation for getting things done. He said the first step was to make the legislature aware of the threat to the Bay so that a strong bill could be passed in a later year. McAteer therefore proposed a legislature study commission, to analyze the issues affecting the Bay and report to the following year's session with specific recommendations.

The McAteer bill was passed, and the senator was named chairman of the Bay study commission. As chairman, McAteer scheduled a whirlwind series of 12 weekly public hearings in all parts of the Bay Area, with statements sought from experts and ordinary citizens alike. Some witnesses testified eloquently to the dangers of unrestricted filling, and others spoke with equal eloquence for increased development of the Bay, meaning filling, because of the economic benefits the bay provides. (Along the way, Berkeley gave up its massive fill plans.)

In slightly less than four months the McAteer commission had agreed that it would recommend the establishment of a successor commission to: (1) make a more detailed that study of the Bay than the McAteer commission had had time for, (2) use the results of that study to prepare a plan "for the conservation of the waters of the Bay and the development of its shoreline," and (3) of great importance, protect the Bay from further piecemeal filling while the plan was being completed; the new commission would thus have veto power over new filling and dredging for a three-year planning period.

"San Francisco Bay is the greatest single natural asset of the region," said the McAteer commission in its report. "The public interest requires creation of a governmental mechanism to balance competing interests in the Bay, and to weigh all the alternatives in making choices relative to the Bay."

Could any legislator—even as skilled a legislator as McAteer—steer such a bill through the California legislature, which in 1969, had shown only sporadic interest in planning and conservation? McAteer vowed to try. The Save San Francisco Bay Association, the League of Women Voters, and many other citizens' groups pledged their help.

And then a disc jockey named Don Sherwood, whose 6 to 9 A.M. radio program was one of the most popular in the Bay Area, joined the fight. "Write your legislators and tell them what the Bay means to you," said Sherwood to his morning audience, and write his listeners did. Telegrams and letters by the sackful arrived on the desks of surprised legislators, more mail than was sent on any other subject in the 1965 legislature. Some citizens did more than write—they bombarded legislators who were thought to be hostile to the McAteer bill with sacks of sand that carried notes reading, "You'll wonder where the water went if you fill the Bay with sediment."

McAteer and the citizens prevailed. The Bay Conservation and Development Commission came into existence in late 1965, with the three assignments McAteer had fought for: to make studies of the Bay, to plan for the Bay, and to protect the Bay from unnecessary filling while the plan was being prepared. The legislature thus specified that any individual or any governmental agency wanting to place fill in the Bay or to dredge in the Bay had to first obtain a permit from the BCDC, and the BCDC was to hold public hearings before deciding whether to grant a permit.

The BCDC Plan

The 27-member BCDC completed its work on time. Its Plan for San Francisco Bay was submitted to the 1969 Legislature, with this summary:

> San Francisco Bay is an irreplaceable gift of nature that man can either abuse and ultimately destroy—or improve and protect for future generations.
>
> The Bay can serve human needs to be a much greater degree than it does today. The Bay can play an increasing role as a major world port. Around its shores, many job-producing new industries can be developed. And new parks, marinas, beaches, and fishing piers can provide close-to-home recreation for the Bay Area's increasing population.
>
> But the Bay must be protected from needless and gradual destruction. The Bay should no longer be treated as ordinary real estate, available to be filled with sand or dirt to create new land. Rather, the Bay should be regarded as the most valuable natural asset of the entire Bay region, a body of water that benefits not only the residents of the Bay Area but all California and indeed the nation.

Harold Gilliam, environmental writer for *The San Francisco Chronicle*, described the Bay Plan as "a Magna Carta for the Bay—a declaration of the Bay's right to live." And the Bay Plan, together with BCDC's methods of public involvement in the planning, won a top merit award from the American Institute of Planners.

Whether the plan should be carried out, and whether BCDC should continue to exist, were among the top issues of the 1969 legislature. As the result of great citizen support, similar to that of the 1965 campaign, a strong bill was passed, and BCDC was directed to continue its work, with somewhat broader responsibilities. As a result of the 1969 legislation, BCDC has three major duties:

1. To regulate all filling and dredging in San Francisco Bay (including San Pablo and Suisun bays, and all sloughs that are parts of the Bay system) in accordance with the law and the Commission's Bay Plan.
2. To have limited jurisdiction over substantial developments within a 100-ft strip inland from the Bay. Within this strip, the Commission's responsibility is twofold: (*a*) to require public access to the Bay to the maximum extent feasible, consistent with the nature of new shoreline developments, and (*b*) to

try to ensure that existing shoreline property suitable for high-priority purposes such as ports, water-related industry, and water-related recreation, is used for these purposes, thus minimizing pressures to fill the Bay.

3. To have limited jurisdiction over any proposed filling of salt ponds or managed wetlands (areas diked off from the Bay and used for salt production, duck-hunting preserves, etc.). These areas, although not subject to the tides of the Bay, provide wildlife habitat and also provide water surface important to the climate of the Bay Area. If filling of these areas is proposed, the commission is to encourage dedication or public purchase to retain water area. And if development is authorized, the commission is to ensure that such development provides public access to the Bay and retains the maximum amount of water surface consistent with the nature of the development.

Lessons From the BCDC Experiment

1. Obviously, many of the problems of our physical environment can be solved only on a regional basis. A consensus was achieved in the Bay Area that the Bay could be protected only if it were treated as a single body of water, not as divided waters to be controlled separately by 9 counties and 32 cities. This aspect was discussed by Russell E. Train, former Chairman of the Council on Environmental Quality,* in his foreword to *The Saving of San Francisco Bay*, which was published last year by the Conservation Foundation in Washington, D.C. Train wrote:

> The BCDC story has [the] important theme . . . of regionalism, of adjusting the new public powers to a community large and inclusive enough to have both the perspective and the incentive to see the Bay as it really is. . . . [A]esthetic or natural resources vital to a regional population cannot be adequately planned for and controlled solely by governments who see only their small share of them, and who are under heavy pressure to capture the property tax revenues intense development could bring.
>
> The Council on Environmental Quality seeks reform of the regulatory structure over land-use decisions that have greater than local impact. We believe that the states, as a condition of federal assistance, should be called upon to control their areas of critical environmental concern, such as coastal and estuarine areas and scenic shorelands, among others.
>
> Some have purported to see in this approach a dire threat to cherished values of municipal home rule and local autonomy. And so it is, if one regards the power of a relatively small community to make the major decisions affecting the interests of regional populations as a proper adjunct of home rule. But such an approach does leave inherently local land-use issues with local governments.
>
> However, like the nineteenth century pioneer ethic towards the land, unthinking obeisance to notions such as home rule can be an obsolete reflex in a more complicated

* Now Administrator of U.S. Environmental Protection Agency.

time. The acceptance of a legitimate state and regional interest in the uses of San Francisco Bay and its shorelands by the local governments in whose jurisdiction they lie is evidence of a growing maturity in both intergovernmental relations and governmental policy.

The BCDC experience is a very hopeful one. Concerned citizens demonstrated a vigor of purpose and a tenacity that outlasted setbacks, and persisted year after year and session after session until the legislature responded. It is as though, having come to the end of a long westward journey of conquering the land, Californians contemplated the Bay and resolved to let it live.

2. Regional planning and regional management of regionally important resources need not mean, and should not mean, riding roughshod over local governments. One way to prevent this from happening is to design the regional agency so that local government (and other levels of government) are fully represented. This was done in establishing BCDC, which has a membership of 27, as follows: 7 representatives of the public (5, including the chairman and the vice-chairman, appointed by the governor and one each by the State Senate Rules Committee and the Speaker of the Assembly); 9 county representatives, consisting of one member of the Board of Supervisors of each of the 9 Bayfront countries; 4 representatives of cities, appointed by the Association of Bay Area Governments; 5 representatives of state agencies such as the State Lands Commission, the Resources Agency, and the Water Quality Control Board for the Bay Area; and two federal representatives, one from the Army Corps of Engineers and the other from the Environmental Protection Agency. Although a 27-member commission is large and could become unwieldy, the BCDC experience certainly shows that a body with this broad and varied makeup can effectively plan for and manage regional resources.

3. To return to Mel Scott's theme, why do so many people so highly value the Bay? Why was it possible to take governmental action to protect the Bay? The answer is more complicated than enjoyment of Bay views, or protection of fish and wildlife, or simply preservation of a body of water that is, after all, only valuable, not sacred.

I think the answer lies partly in a growing belief that man's management of nature need not be synonymous with destruction, that indeed man's intervention may be necessary to maintain what nature has provided. Coastal estuaries such as San Francisco Bay are not, in a state of nature, permanent and unchanged. Dr. Donald W. Pritchard, director of the Chesapeake Bay Institute, put it this way:

> If the people packed up and left the Chesapeake Bay alone with its shellfish and finfish and plankton and birds and beautiful sky and rugged waters and marshland and all the wild creatures that live in its environs; if the Bay were allowed to return to nature 100 percent—in the course of some hundreds of years it would cease to exist.

Siltation and erosion would fill it in and it would become the world's largest mud pie. . . . We are living in what could be called the Age of Estuaries. For *x* number of centuries there were no estuaries. Now there are estuaries, primarily because of the glaciers. If there were no human beings to maintain them, in *x* number of centuries there would be no estuaries.

I think further that San Francisco Bay became to many people a symbol of the environmental questions faced everywhere in America. Do we of the present generation have the right to destroy irreplaceable resources? Do we have the right to shrink a magnificent bay into little more than a river, forever depriving future generations of the enjoyment of it? (The BCDC Bay Plan states as one of its goals "guaranteeing to future generations their rightful heritage from the present generation: San Francisco Bay maintained and enhanced as a magnificent body of water that helps sustain the economy of the western United States, provides great opportunities for recreation, moderates the climate, combats air pollution, nourishes fish and wildfowl, affords scenic enjoyment, and in countless other ways helps to enrich man's life.")

Moreover, if we cannot manage to protect San Francisco Bay, how can we manage to solve the other and more difficult problems of modern America? And, perhaps most important of all, can ordinary citizens, working through the American political system, make their voices heard and manage to achieve a healthy balance of conservation and development?

In the final paragraph of *The San Francisco Bay Area: A Metropolis in Perspective*, Mel Scott wrote:

> Because the San Francisco Bay Area has been more generously endowed with natural beauty than most of the large urban areas of the world, its residents have an especially weighty stewardship to discharge. Not for ourselves and our own posterity but for the global community should we strive to safeguard the beauties of this magnificent combination of bays and islands, mountains and valleys—and to enhance the heritage received from nature by skillfully wedding the structures and roadways of the evolving regional metropolis to the physical environment. To achieve so lofty a goal we must do no less than perceive new obligations of the individual to society, fashion new laws for controlling individual action in the interest of the entire community, devise a workable system of metropolitan government, and weld into a new creative force the talents of all those concerned with metropolitan regional development—elected officials, businessmen, industrialists, and their employees, members of the professions, and workers in the many governments of the San Francisco Bay Area. Are we equal to the challenge?

Local Government—Will It Survive?

ROBERT HORTON

As the United States of America prepares to enter the third century of its existence as a nation, we face an internal crisis in federalism. We need to make up our minds: are we going to restructure and strengthen the federal system, or are we going to abandon it? During the last quarter-century, we have experienced extremely rapid urbanization; now more than 70 percent of our population resides in less than 3 percent of the land area. During this period, governmental trends have run in both directions. In some areas problems have become so great that tendencies have been to shift functions to higher levels of government and to move closer and closer to a unitary form, running from national headquarters in Washington.

The Nixon administration, through grant decentralization and general revenue sharing, has offered us a chance to prove that the states and local government can make the "new federalism" work. However, they wrapped up 50 years of unresolved problems and threw them over the fence into our playing field.

We have slightly more than 3000 counties in the United States. Most areas started out as geographic subunits of the state for administrative purposes. Now, however, they are only part of an administrative structure at the local level that contains more than 85,000 units of governmental or quasi-governmental entities. The states have had a basic responsibility for organizing the local government delivery system and have permitted so great a fragmentation of this system that reorganization and reform are imperative if we are to main-

Mr. Horton is the Director of the Environmental Planning and Management Project in the Office of the Mayor, Metropolitan Government of Nashville-Davidson County.

tain an effective participatory role at the local level in planning, programming, and implementing services in rural or urban areas.

We are living in an era so different from the last generation's that we do not yet recognize the magnitude and nature of the changes that have occurred in our country. When we talk of the problems of our urban society of education, of politics, of the world of work, of environment, we must place the issues within the context of the changes that have occurred and are continuing within our borders. We are entering an era of change during the next 20 years in which the variety of problems facing us is more than man has faced within the past century. Automated machines and modern technology confront us with powerful agents of social change. The Industrial Revolution and the railroad, which changed the economics of the United States, the character of life, and the environment, also created many problems; many are still with us. Now we have a modern technology: machines can handle and understand information, communicate it, and do things with it. But as we move into a knowledge-centered society; do we know how to care for and manage the knowledged worker or do we know how to organize local government system to plan and deliver the services necessary for the health, comfort, safety and growth and productive capacity of this worker?

We live in a very large country where the application of technology varies tremendously. Are we going to reform and design a governmental structure that will permit areas to move into a technological world with some control for their own destiny? I feel, for example, that the issue of environment requires strong knowledgeable inputs on a day-to-day basis from the local level. True multilevel decision making is necessary. Of course the state and federal governments have their roles and responsibilities.

To me, the key issue in the preservation of the federal system lies in creation of strong, efficient units of local government having both the areas of jurisdiction and the authority to meet the needs of our sprawling urban centers.

I am not suggesting that you should follow the example of Nashville–Davidson County. State laws and state constitutions differ, and any local government reform should be developed and designed to fit the problems within the respective state or region. I do believe, however, that a discussion of the events in Metropolitan Nashville–Davidson County occurring over the last 21 years does point to one approach that should be seriously considered or evaluated in many areas as local government reform is being studied. I also believe that the organizations formed, the directions tried, and the avenues from which support came, can help others plan the campaign for metropolitan government before it is waged.

The Nashville–Davidson County Story*

The Nashville area was caught in the same population explosion as the rest of the country, with the same resulting difficulties, including the disproportionate burden of the obligation and costs of government falling on the city, and the inability of a rural-type county government to serve the urban needs of suburbia.

In 1951 the Tennessee Taxpayers' Association issued a report recommending one completely consolidated unit of local government for Nashville–Davidson County.[1]

The Nashville and the Davidson County Planning Commissions, cooperating with each other, studied consolidation. The solution then contemplated was for the city to annex the urban communities adjoining its boundaries and for consolidation of those functions which were of county-wide benefit and shifting them to the county exclusively.[2] Local political jealousy between the competing city and county systems and administrators frustrated this.

Seeking a nonpolitical solution, the Executive Director of the Davidson County Planning Commission interested 15 prominent citizens in serving on a Community Services Commission to study governments—to analyze, criticize, and recommend changes.[3]

The 1951 General Assembly of Tennessee, naming the 15 commissioners, created the Community Services Commission for Davidson County and the City of Nashville. In June 1952 the commission issued a report, a searching and frank criticism of each area of a local government by its citizens. Its recommendations included "Annexation of suburban Nashville, County responsibility for county-wide functions, city and county home rule, redistricting of Davidson County."[4] Again jealousies blocked implementation.

In the meantime, legislative approval of a constitutional amendment permitting the consolidation of city and county governments was obtained and ratified in a statewide referendum (November 3, 1953). The constitutional amendment was not self-executing because it authorized the General Assembly, which meets biennially in odd-numbered years, to provide for the consolidation of the functions of city and county governments, and until the legislature enacted the necessary laws, nothing could be done.

When the 1955 General Assembly met, the Tennessee Municipal League was successful in its campaign to liberalize the laws under which municipalities could annex territory; no longer was the consent of the people being annexed needed.[5]

* Editor's additions from a 1966 speech of Seymour Samuels, Jr., Deputy Metropolitan Attorney, Nashville.

After the enactment of this statute, three new small cities were formed in 1957, 1958, and 1959 to avoid annexation by Nashville and also to be protected from "metropolitan government."

In response to the request of the Nashville Area Chamber of Commerce, the city and county planning commissions were asked to find a way to improve the local community services.[6] Their report, submitted October 30, 1956, recommended "The Creation of a Single Metropolitan Government to Replace Existing City and County Governments. . . ."[7] The enabling act was then drafted and passed, implementing the consolidation amendment, providing for the consolidation of city and county functions under a new metropolitan form of government, and providing for the drafting of a charter. The Nashville City Council and the Quarterly County Court of Davidson County had already passed companion resolutions authorizing the establishment of a ten-man Charter Commission. On June 17, 1958, the charter was submitted to a referendum of the voters in the City of Nashville and in that portion of Davidson County lying outside the City of Nashville. It was defeated by county voters.

On July 16, 1958, the Nashville City Council annexed an area of 6.91 square miles—4587 people; on April 29, 1960, it annexed 42.46 square miles—82,512 people.[8]

City property taxes added to the county property taxes caused the annexed area's taxes to double, and the city did not provide services. The right political climate to produce change existed.

The Davidson County delegation to the 1961 General Assembly accomplished the following legislative enactments:

1. An Amendment to the metropolitan enabling act, permitting a Charter Commission to be created by a Private Act of the Tennessee General Assembly.
2. An amendment to the Tennessee Code which provided that any annexation of more than one-fourth (¼) square mile or over 500 persons must include an adopted plan of service and a projected time table for extending the service.
3. Chapter 408 of the Private Acts of 1961 creating a Metropolitan Government Charter Commission for Davidson County, subject to approval by vote of the people. This was approved in an election on August 17, 1961.[9]

The charter, prepared by the Charter Commission, was approved in election on June 28, 1962, and the government became effective April 1, 1963.[10]

The 1962 campaign for metropolitan government, was joined by many civic organizations, including the Nashville Area Chamber of Commerce, the Junior Chamber, the League of Women Voters, the Council of Jewish Women, the Federated Business and Professional Women, and others. A Citizens' Committee for Better Government coordinated all activities and conducted a door-to-door campaign for the charter, utilizing the members and the staffs of all the other groups. *The Nashville Tennesseean*, one of the two daily newspapers in the county, adopted the cause of metropolitan government as its own and went all out for the charter.[11]

The charter did not eliminate the satellite cities; it continued their existence without interruption. This was done under the provisions of a general law that precluded the incorporation of municipalities in counties having a metropolitan form of government, unless the charter of such government specifically permits additional cities.[12]

Metropolitan Government Organization

The Two Service Districts Concept

The problem of the smaller cities and their relationship to the metropolitan government, was resolved by dividing the metropolitan government into two taxing districts. The General Services District is composed of the entire county. The Urban Services District is that portion of the General Services District which is within the area that formerly was called the city of Nashville. Separate budgets must be adopted for each services district, and taxes are levied in each of them to support the respective budgets.

The functions of the General Services District are: general administration, police, courts, jails, assessment for taxation, health, welfare, hospitals, housing for the aged, streets and roads, traffic, schools, parks and recreation, libraries, auditorium, airport, public housing, urban redevelopment, urban renewal, planning, electricity, transit, building and building trades codes, housing code, refuse disposal, beer supervision, and taxicab regulation.

In the Urban Services District, functions are additional police protection, fire protection, water, sanitary sewers, storm sewers, street lighting, street cleaning, refuse collections, and wine and whiskey supervision.*

Consolidation

It was decided to reduce the city and county's 87 boards, commissions, departments and agencies to 8 departments and 17 major boards, minimizing duplications. Key consolidations were: Schools, Public Works, Police, Water and Sewer, Codes, Hospitals, General Services.

Strong Mayor–Council Form

The metropolitan government featured a council of 41 (Vice-Mayor, 35 district representatives; 5 representatives at large).

* End editor's additions.

Advantages of Consolidation

Among the numerous advantages of consolidation were the following:

1. Fixed countywide jurisdiction and authority of a municipal type local government with charter authority.
2. United business and civil leadership in the community.
3. Produced a renewed interest and participation of business community in the affairs of local government.
4. Consolidation produced savings and increased efficiency.
   (*a*) Eliminated duplication.
   (*b*) Consolidated fiscal management permits more than $1 million interest on idle fund investments ($600,000 more than old city–county management).
   (*c*) Consolidated purchasing—motor pool.
   (*d*) Savings from consolidated water-sewage program, eliminating small disposal plants and inadequately sized mains.
   (*e*) Savings from data processing, eliminating routine paper work.
   (*f*) The savings in the central housekeeping in general government area. We spent less for these functions during fiscal 1967–1968 than we spent over the old two governments prior to the consolidation.
   (*g*) The consolidation has eliminated approximately 800 jobs in the earlier governments. This does not mean that the total personnel of the government has been reduced, however; many employees were transferred because the new government was committed to extend park and recreation and library services. Also, major increases have occurred in Police, Fire, Public Works, and Water and Sewer personnel, because of expansions and upgrading of these services. Other areas, such as Hospitals, have required additional personnel to reduce the work week from 72 hours to 40. The net effect of the consolidation has been to reduce administrative costs and to make possible the reallocation of these tax revenues to service functions.
   (*h*) Savings from the use of equipment to reduce labor needs. In the Public Works–Water and Sewer Department and in several other areas, the consolidation has made possible the development of work programs sufficiently large to justify the purchase of labor-saving equipment for construction, street work, street cleaning and related fields, whereby some of the cost-cutting techniques that have been used by businesses for years are now being applied to our government operation.
   (*i*) The savings through the joint use of facilities. Our school board and our park board are now making joint purchases of sites, and the school

buildings are so constructed that they may be used as community recreation centers on a year-round basis. School sites 10 feet from the school building are under the supervision of the park board for recreation purposes.

(*j*) The consolidation of Public Works functions has made for better programming of the overall Interstate Highway System in relation to urban renewal projects and programs and to the general development of the community.

(*k*) Having a single tax assessor has led to a much improved assessment staff, and we are now developing a complete tax reappraisal program based on the new property maps and materials that have been prepared in the last three years.

(*l*) The consolidated Police Department has eliminated graft of the old system with improved recruiting procedures. Substantial improvements have been made in police training, following recommendations of the International Police Chiefs' survey.

(*m*) The merging of the two school systems was a major undertaking, and teacher salary equalization, improved curriculum, and a more effective school system have resulted from the consolidation.

(*n*) The Water and Sewer Program, under the consolidation, has sewered more residential areas in the last 4 years than were sewered in the total history of the City of Nashville prior to the consolidation. More than 40 square miles of area either has been sewered or is under contract.

An Evaluation of Metropolitan Government

We have not achieved utopia in Nashville; many problems remain to be solved. Our slogan is, "Let us keep up while we catch up." The backlog of service needs that accumulated over the last 50 years, when combined with the spectacular growth in urban development and population that we have experienced in the last four years, has created problems. The rising level of public expectation for solutions now to the problems of long duration has further complicated our efforts. Inflation and the rising wage levels in our community, resulting from an unemployment rate of 3 percent or below, have made it necessary to raise salary levels for employees in order that we may compete in the labor market.

All the foregoing factors have added to our costs and have added to the pressures that we in Nashville, along with the other urban centers of the United States, have experienced. Overzealous supporters of consolidation may have led some citizens to believe that all problems would be solved immediately after the consolidation and that taxes could be lowered. This was not the viewpoint of the knowledgeable supporters of the consolidation. The taxes have gone up in certain categories because service needs had to be met on a community-wide basis. Some people who had not paid a fair share of taxes for facilities they were using, have been and are paying for these services. We have doubled the school budget in four years because more funds were needed. We have built parks and libraries and roads because they were needed.

The taxpayer, I believe, has gotten more for his money in direct services than would have been the case had the former City of Nashville and Davidson County continued to duplicate services and squabble over schools and revenue allocations. Nashville–Davidson County has a low effective tax rate (4.11 per $100 assessment in the General Services District + 1.89 additional in the Urban Services District).

The Metropolitan Government of Nashville–Davidson County provides a framework of local government through which the problems of a growing community may be resolved, services rendered to people, and community-wide facilities constructed on a planned basis to promote economic growth and to improve the urban environment for all the citizens of the area. We have not been able to make the concept of taxpaying popular, but we have given the taxpayer more service for his tax dollar.

References

1. Tennessee Taxpayers' Association, *Report on a Detailed Survey of the Financial Condition of Davidson County, Tennessee, with Recommendations*, Nashville, October 1951, Section IV, p. iv.
2. Personal Interview, May 15, 1966, with Charles W. Hawkins, formerly Executive Director, Davidson County and Metropolitan Planning Commission.
3. *Ibid.*
4. "A Future for Nashville," A Report of the Community Services Commission for Davidson County and the City of Nashville, Nashville, June 1952.
5. Chapter 113, Public Acts of Tennessee, 1955.
6. Personal interview, May 15, 1962, with Charles W. Hawkins, formerly Executive Director, the Davidson County and the Metropolitan Planning Commissions. Also: Brett William Hawkins, "Sources of Opposition and Support for Metropolitan Reorganization: The Nashville Experience," Ph.D. thesis, Graduate Faculty of Vanderbilt University, Nashville, August 1964, p. 63.
7. *Local Government*, Fact Book Series, Vol. II, Metropolitan Planning Commission, Nashville, 1966, p. 2.
8. *Ibid.*, p. 3.
9. *Loc. cit.*
10. *Ibid.*, p. 4.
11. Brett William Hawkins, *op. cit.*, pp. 142–145, 169–176.
12. Tennessee Code Annotated, § 6-3716.

Organizing for Environmental Management

WILLIAM FELLING

The programs I shall discuss represent an attempt by the Resources and Environment Program of the Ford Foundation to help test the validity of the regional concept of environmental management. I would like to discuss the rationale that led us to try these experiments, as well as some of the criteria evolved after extensive site visits to most metropolitan regions in North America and to briefly describe the six grants that have been made. Finally, I would like to share some of the lessons we and our grantees have learned.

The Rationale

During the last decade it has become abundantly clear that a major cost of urbanization is the accumulation and concentration of increasingly intractable wastes. Mostly, however, the problem has been seen in terms of its symptoms: dirty air, befouled and contaminated rivers, endangered water supplies, and precious lands preempted for refuse; remedies have been directed at the sore points as though each were a separate issue. Various levels of government operate separate agencies and programs for air pollution control, water supply, sewage treatment, and garbage and trash disposal. This way of slicing the problem is historically understandable but is neither logical nor in our crowded world effective.

Pollution is nothing more than the presence of materials where they are not wanted. The pollution that today chiefly concerns us is that which results from the discard in one way or another of the residuals of production or consump-

Dr. Felling is the Program Officer of Resources and Environment for the Ford Foundation.

tion. With regard to that pollution we have a simple, ineluctable choice: either reduce the residuals (by reuse, e.g.) or decide what to pollute. Clearly governments so divided that some people are charged with keeping the air clean, others with keeping the waters clean, and still others with getting the trash out of the way have no single control over the basic choice between reducing wastes or disposing of them and the secondary choice of where to put them. Obvious as is this disparity between the nature of the waste management problems and the administrative machinery established to solve it, too little serious effort has so far been made to do better.

The move to regroup traditional divisions between governments in metropolitan areas and among departments of government to help cope with urban wastes has been an attempt to put all the "housekeepers" under one roof, rather than a recognition that waste management is a single problem. In the federal government formerly scattered pollution control agencies that assist cities have been assembled under the new Environmental Protection Agency. Many state governments have similarly repackaged the old "environmental" agencies. So have some municipalities. At every level, however, it has proved easier to set up an overarching bureaucratic tent than to effect genuine integration of services. The inability of New York City's Environmental Protection Administration to diminish the autonomy of the departments under its jurisdiction is typical. In the federal EPA the old programs for controlling air pollution, building water treatment plants, and assisting in the disposal of solid wastes remain distinct under the administration of agencies whose missions have not obviously been broadened. Drawing new organization charts or bringing in new bosses has proved insufficient to overcome separatism enshrined in vested interests.

Two and one-half years ago staff assisted by consultants reviewed the waste handling practices and environmental management capacities of a large number of metropolitan areas to see whether in some cases, with Foundation help, the effective obstacles might be overcome. Essentially we sought out places in which the typical difficulties with entrenched bureaucracies, jurisdictional conflict, and public indifference were least, and able and suitably placed leaders were ready to try for a change. In such situations private funding for a short period might enable selected regions to develop the methods, the staff competence, and the public support needed to plan and implement regional environmental management.

Criteria

Among the criteria that made some urban regions attractive as places to try experiments in regional environmental management, the following are central:

1. The metropolitan region is large enough to present a representative array of problems and of administrative complexities, but small enough to make

possible innovative personal leadership and to mount an effective attack. This requirement turned out to put us in urban population ranges of 150,000 to 1.5 million people. The largest area has a population of 1.3 million; the smallest, 250,000. Since by and large these medium-sized areas are also those in which the maximum population increases are taking place, they face the most difficult problems of planning for growth.

2. The waste management problems of the areas are sufficiently self-contained that control can be exercised by some level of regional government.

3. The existence of regional structure with the potential to govern effectively. This has been found most often in counties containing one predominant urban center, not as we had thought might be the case in regions with strong Councils of Government. New England, in this regard, is an exception, and a different mechanism would have to be used if experiments are to be tried there.

4. A particular, well-defined environmental problem exists on which planning and administrative energies can be immediately brought to bear and methods having general application tried out under the pressure to resolve one urgent difficulty.

5. Political and other complications that grant aid cannot address are absent or minimal.

6. Most important of all, there must be at least one administrator in a key position who has the demonstrated ability and drive to get things done.

The Grants

Six grants totaling $2.65 million have now been made. Grantees include the San Diego County Board of Supervisors, the Commissioners of Erie County (Pennsylvania), the Metropolitan Government of Nashville–Davidson County, the Municipality of Metropolitan Seattle (METRO), the University of British Columbia (in Vancouver), and the University of Hawaii. All but two of these grants have a two-year term and will come up for a renewal decision within the next year. (See list at end of chapter.)

In varying degrees each grant seeks to help local and regional officials to strengthen and expand the environmental data base, to improve the capacity of elected officials and their civil servants to anticipate the possible consequences of the variety of environmental decisions they must face, and to provide for a more effective flow of information to and from the broadest possible range of individuals and organized citizens, so that environmental decision making can become truly participatory and more informed.

In the early efforts in Hawaii, Vancouver, and San Diego, we were asked to provide support for a more complete data base and better management and simulation models in such areas as land use planning, water and waste management, transportation, and pollution abatement. These localities are only now beginning to respond to the challenge of packaging their results into a more useful form for their several publics. In fact, our grant in Hawaii is entirely to support the creation of a Citizens' Advisory Committee chaired by a state senator. The technical content, such as data gathering and model building, is being funded by a matching National Science Foundation (NSF) grant. Similarly, the grant to METRO supports a Citizens' Task Force, organized to enhance both the quality and the quantity of citizen participation in the regional environmental planning efforts. In this grant the Seattle District Office of the Corps of Engineers has a central and imaginative role.

Understandably, the extreme complexity of the interinstitutional participation being attempted has brought about major internal realignments within several of the grants, notably in San Diego, Vancouver, and Hawaii. All now have realigned their management and redefined institutional ties to local and regional governmental agencies, displaying increasing concern with "how and when to go public" with the data and tools being generated by these projects. A brief description of the projects follows.

San Diego County

One of our earliest visits was with elected officials and staff in San Diego County. There momentum was developing to provide services on a regional basis (planning, sewage treatment, transportation), and the county government has just been restructured in response to this momentum. There was a strong feeling that the county should not allow itself to succumb to the unbridled push for urban growth. We were asked to help a newly created Environmental Development Agency function truly differently from the way its predecessor departments had operated. The project has survived a switch in political leadership and major internal shifting, all of which has been sufficiently important to require a reexamination of goals for the Integrated Regional Environmental Management (IREM) project.

Hawaii Environmental Simulation Laboratory

During the same time we were struck by the unusual attention paid to environmental issues in the local papers on the island of Oahu. The advantages of studying the state of Hawaii as a nearly closed ecosystem had occurred to many people, but no one knew quite where to start. Faculty from the University of Hawaii, together with representatives from state and local government and the Oceanic Institute, proposed the creation of the Hawaii Environmental Simulation Laboratory (HESL), to begin to develop sufficient information to anticipate the consequences of various mixes of its three major economies—military, tourism, and agriculture. HESL began with a detailed study of the land developments around Kaneohe Bay and their impact on the rapidly deteriorating quality of the bay. It has now moved into transportation and hydrological studies and hopes to eventually have both island-wide and statewide environmental management inputs. This experiment was the first in which NSF and the Ford Foundation provided joint support—NSF for the development of better data and models, Ford for the activities of the Citizens' Advisory Committee.

University of British Columbia

The Inter-Institutional Policy Simulation project (IIPS) in Vancouver was not conceived as one of the Foundation's experiments in regional environmental management, although much of what followed came out of our earlier discussions with the many groups and agencies participating in the IIPS project.

Originally conceived to involve professionals from city and regional government, as well as faculty and students from the university, IIPS now is also receiving active support from both the provincial and the federal government. As in San Diego, the people of the Greater Vancouver Regional District have become increasingly concerned with future quality of their environment. Aware that they will be (and already have been) asked to make some major decisions regarding development and growth in the region, and that the public must be involved all along the way, members of the IIPS project see it as one of the ways of improving and broadening the decision-making processes. Of the six experiments receiving support from the Foundation, IIPS undoubtedly has the heaviest modeling and simulation content at this time.

Erie County

Federal legislation has made the states responsible for setting and enforcing environmental standards. Even though states have neither the machinery nor the capability to monitor effectively on a statewide basis, they are reluctant to relinquish their authority to local governments except when local government clearly has the capability, as is the case only in our largest metropolitan areas. In Erie County we were asked to help strengthen an existing county health department so that it could act with the authority of the state in county pollution control and abatement problems. To date, the Pennsylvania Department of Environmental Resources has given the Erie County Health Department responsibility for landfill monitoring, water quality, environmental health, and the monitoring of the lake around Erie, and may soon also transfer to it responsibility for air pollution monitoring and control. The success of this project is the direct result of an unusually confident working relationship between the director of the Department of Environmental Resources and the project director in Erie. The final test will be the state's ability to replicate this program in other counties.

Nashville–Davidson County

The initial focus of the Nashville–Davidson County grant is the design and implementation of a county-wide solid waste collection and disposal system. Despite the consolidation of city and county government a decade ago, functional consolidation has been slow to take place. The management of solid waste, for example, lies within the jurisdiction of four metropolitan government departments. This project is designed to weld the directors of four departments into a management team, giving them special training in environmental management and assisting them in mobilizing the resources within their agencies for a series of special studies, to draw in expertise from the region's 14 colleges and universities, and to involve both the business community and private citizens in the management process. The grant is so recent that no progress report can be meaningful.

Municipality of Metropolitan Seattle

Our most recent grant has been to the Municipality of Metropolitan Seattle (METRO). Created more than a decade ago, METRO was voted the power to perform one or more of six functions: sewage disposal, public transportation, comprehensive planning, water supply, garbage disposal, and park administration. Its earliest and most dramatic success was the cleaning up of Lake Washington. More recently it has moved into public transportation and comprehensive planning. The District Corps of Engineers is conducting a regional storm-water runoff study, and EPA has contracted for a regional waste management study. The cooperation between the Corps, local and regional planning agencies, and citizens organizations, led them to seek our support to try to institutionalize this relationship through the creation of a Citizens' Task Force. The task force has its own staff, access to all the technical planning and studies under way, and a budget, essentially matched by Corps money, to package this information into digestible bites for a variety of citizens' groups.

Some Lessons Learned

In all these projects the grantees see the issue of environmental management as a vehicle to encourage wider participation in decision making and to help strengthen the interinstitutional connections that are necessary to deal with regional problems. Starting with an appointed administrator selected for his imagination, integrity, and position, they have attempted to develop new coalitions of people in government, education, and industry, as well as private citizens. The resulting institutional arrangements are fragile though promising, and they require special attention. The mix of elected officials, civil servants, educators, and the public is at times abrasive; the integrated approach to management is untested and easily turned away; the project directors must sometimes make decisions of unprecedented (for them) importance. Thus it is not surprising that all the project directors have asked that the Foundation maintain a particularly close relation and that it stand ready with professional advice.

Early in 1972 the officers of the Foundation approved an internal budget of $150,000 to monitor and provide technical assistance to these projects. These funds are being used in several ways to provide technical and organizational assistance to the grantees and to initiate a program to strengthen the communication links between them. Representatives from all the projects met twice within seven months to familiarize themselves with one another's projects, to exchange technical notes, to discuss organizational and institutional arrangements that have worked or failed, and to advise Foundation staff and consultants on how they can be of greatest use. At one of these meetings representatives from NSF, EPA, the Rockefeller Foundation, and the Canadian Ministry of State for Urban Affairs participated. The notion is that since the grantees have much to learn from one another, continued communication among them will not only help prevent duplication of technical development but will also help produce a synergistic effect from them.

It is safe to say that these attracted the attention of other funding agencies. Although several of the projects have received federal support, as in the case of an NSF grant to Hawaii and EPA's categorical grants to San Diego, Erie, and Nashville, none of these funding agencies are ready, as yet, to provide the kind of broad, interinstitutional support that seems to be called for.

Even in the best of all possible worlds we have realized that there have been substantial differences between the goals and objectives of each project as viewed by the Foundation and the grantees at the time the grants were approved. The need to deal with real-life changing situations has in some instances compounded these differences. An accurate account of what actually happens, particularly among people, as these programs progress, could be useful information for other regions and institutions that might want to try similar experiments. Accordingly, each program has appointed a historian to document major events. Some of these reports may be sufficiently generalizable to warrant widespread distribution; only time will tell.

Eighteen months after the first experiments were funded, it is safe to say that all the people involved have learned a great deal and that some interesting similarities have begun to emerge from projects that at first blush appear to be different.

For the institutions involved, government with its civil servants, academics, citizens' organizations, and industries, the process of seeking a common ground for dealing with mutual problems has been much slower and more painful than anticipated. Governmental agencies have been noticeably reluctant to encourage active participation in environmental decision making by outside citizens' groups. It is only when the two sides try to work together that they begin to realize the differences between each one's concept of the goals and objectives of environmental management. Everyone wants the projects to succeed, but seldom for the same reasons. Academics revel in the environment of free and open access to all information; industries and governments have vested interests they are charged to protect.

From project to project there is no consensus about how the experiments should be managed. The management structure ranges from decisions by the consensus of a large working core group to rulings of a single project director. Each project seems to settle on that structure which makes the participants feel most protected.

The management problem is more complicated for the four grants to regional government, in that some of the civil servants who serve those governments and may be very important to the projects, are vulnerable to changing political situations. Still, if environmental management is basically the business of government, such variables are facts of life that must be dealt with.

In several of the projects the publicity attending the announcement of the grants raised public expectations well beyond the short-term capacity of the programs to respond. The result was a general dampening of the credibility of the experiments in the eyes of the public. The whole issue of when and how to "go public" with the information produced within the projects has yet to be resolved. Vancouver will argue against going public until it is reasonably sure that it can field most requests it might receive for environmental data. Hawaii insists on the importance of recognizing the effectiveness of the process of model development itself as a device for stimulating and facilitating communications between the four participating sectors of society.

From the Foundation's view of the grantees, we have learned that it takes 8 to 12 months before the regional representatives agree on the kind of Foundation help they need and another half-year before tangible results appear. Also there seems to be a direct relationship between the ferment among a project's participants and the extent to which the project is facing real-world situations. Indeed, this ferment has made the Foundation aware that in these instances, unlike some others, it has an obligation after the grants have been made to provide technical assistance to help restore calm.

Foundation assistance has generally taken two forms. On the one hand the grantees need a sympathetic authority to listen to the problems the grants have exacerbated and to help pour oil on troubled waters. Thus the grantees have begun to view the consultants monitoring the grants more as their representatives to the Foundation than as our consultants. Since all these grants have been a mixed blessing to the grantees, this relationship has been encouraged.

The experiments have had a profound impact on the individuals involved.

Two of the project directors have resigned, one was forced to resign. In at least two instances it became clear that the man who was the principal conceptualizer of the project was perhaps not the ideal person for its day-to-day management. Since this had not been anticipated, there was no provision for either continuity or change of leadership. However, as noted earlier, the grants were for short terms, recognizing that because they were experimental, both the Foundation and the grantees might want to make some basic changes before considering extensions. In any case if any of the grants is to be renewed, the new program will look substantially different from the original one.

In most instances these grants were to help nourish new institutional connections, often in situations characterized by little or no previous contact. Here our consultants have been called on to play a broker role and to encourage new alliances. Where the projects have chosen a form of group management, the monitors often have had to help keep peace between dissenting members of the management team.

Remaining Problems

For all these experiments in regional environmental management, several problems remain. It is very difficult to assemble an adequate environmental data base and to develop the analysis needed to anticipate the consequences of a variety of environmental decisions. In addition, there remains the extremely complex problem of translating the results of technical planners and modelers into a language understood by the various decision-making publics. Each distinct public views the environmental issues at a level of abstraction suited to the kinds of decision each must make. What appears to be needed is a human analog to the compiler in computer systems, a group of people capable of translating the planners' jargon into the language of private citizens and with the necessary connections to the public to ensure that planners are aware of the concerns and priorities of that public. It has become increasingly apparent that in attempting to develop regional models, the environmental management obstacles are less technical than human. Interinstitutional communication with a minimum information loss in the loop should be a major objective.

To this end the Foundation has added to the team of project consultants several professionals who specialize in the use of various media to move information into action. In their brief exposure to only two of our projects, they have noted that there is no patent solution to the regional communications problems, which often are being dealt with by people who are technical professionals but amateurs in communications. The communications problems require the same level of professional sophistication as modeling, simulation, and synthesis. How these media experts will be used, if at all, by our grantees has yet to be determined. For now all we can say is that they will be at the next general meeting of our environmental management grantees.

As I said at the outset, the six projects mentioned here represent the result of negotiations that extended over several years. Some may be renewed for a similar two-year term, but we do not anticipate any major new starts, preferring rather to continue to follow these experiments to their logical conclusions. A detailed listing of the grants follows;

1. *The University of British Columbia* ($372,000) project to build an analytical model of the Vancouver region.

2. *County of San Diego Board of Supervisors* ($725,000) partial support of a program to develop a land use model, put it into use by local decision makers, and strengthen community support for rational planning and decision making.

3. *Hawaii University Research Corporation* ($113,000) in partial support of the establishment of the Hawaii Environmental Simulation Laboratory (HESL).*

4. *The Commissioners of Erie County, Pennsylvania* ($297,000) partial support for an experiment in managing waste as a function of regional growth.

5. *Metropolitan Government of Nashville and Davidson County* ($650,380) to develop a comprehensive regional waste management system, to include the related problems of air and water pollution, transportation and land use planning, and to develop an environmental management team in local government.

6. *Municipality of Metropolitan Seattle* (METRO) ($246,400) to stimulate citizen participation in planning a comprehensive environmental management program for the Seattle area.

* *University of Hawaii,* $150,000 grant renewed for partial support of the activities of the Hawaii Environmental Simulation Laboratory.

The Tahoe Experience

CHRISTOPHER DEGENHARDT

History of the Tahoe Regional Planning Agency

Lake Tahoe and the surrounding region are frequently described in superlatives: one of the clearest lakes in the world; one of the largest, highest, deepest, and most beautiful.

Over time, these exceptional and peculiar assets have attracted people to come to Lake Tahoe and enjoy them. During the 1950s gaming casinos were promoted all-year attractions. With the completion of Interstate 80 and the 1960 Winter Olympic Games at Squaw Valley, the Tahoe area emerged as a major recreation destination. These development pressures brought with them increasing public concern about the quantity and quality of development.

The events that trace the response and concern of citizens and public agencies are a record of the evolution of the Tahoe Regional Planning Agency.

Late in the 1950s the Lake Tahoe Area Council (LTAC) was formed. This citizens' group has great concern for the environment and acts as overseer of environmental studies, as well as being a clearing house for information among governmental agencies within the Tahoe Basin. In 1965 another citizens' group was formed; the League to Save Lake Tahoe, too, has become a watchdog of the area's natural and irreplaceable beauty. The Tahoe Regional Planning Commission was created in 1961 along with an advisory committee, and together these bodies were primarily responsible for the preparation of the 1980 Tahoe Regional Plan. The plan was never approved or adopted by all the participating agencies.

In 1964 the Lake Tahoe Area Council and the Tahoe Regional Planning Commission pushed for a cooperative study by the states of Nevada and California to investigate the feasibility of a regional planning agency. Sub-

Mr. Degenhardt is Vice-President of Eckbo, Dean, Austin & Williams, Consultants to the Tahoe Regional Planning Authority.

sequently, the legislatures of California and Nevada created the Lake Tahoe Joint Study Committee. The committee recommended the formation of the Tahoe Regional Planning Agency representing both state and local governments, which made it necessary to set up interim agencies under essentially identical state laws pending the consent of Congress. The California legislature created the California Regional Planning Agency in August 1967, and its first meeting was held in November of that year. The parallel Nevada agency act was passed in late 1968 and became effective in early 1969. These agencies, under their respective state laws, acted as much as possible as if they were joined together by a binding compact. Each agency had a Technical Advisory Committee to review, study, and recommend matters that should come before the monthly meetings of the governing bodies of the respective agencies. Each agency had parallel state-derived jurisdictional authority over its respective county, city, and other local organizations in the Lake Tahoe Basin. Indeed, it might well be argued that the first attempts at truly regional management occurred. There was no binding accord between the two states, but significant cooperative efforts were made with authority over local jurisdictions.

Based on the California and Nevada companion state laws, Congress passed, with small amendment, a Senate bill that was signed by the president on December 18, 1969, as Public Law 91-148. This act was approved by the governors of California and Nevada on March 17, 1970, and the organization meeting of the new agency was held on March 19. Except for the enactment clause and six additions, the language of the act is directly quoted from the California and Nevada joint acts.

The six additions include:

- Appointment by the president of a nonvoting "representative of the United States."
- Authorization by the secretaries of the Interior and Agriculture to cooperate with the agency as compatible with "normal duties" of their departments.
- That any additional powers conferred on the agency by either state must be consented to by Congress.
- That nothing in the act shall affect the powers, rights, obligations, or applicability of any law or regulation of the United States regarding the region or waters of the Compact area.

The Compact Act establishes the legal basis for planning control or "regional management" in the Tahoe Basin, and as such, it gives the Tahoe Regional Planning Agency (TRPA) greater authority than perhaps any other regional agency in the states.

The issues that surrounded the establishment of the TRPA and the granting of such far-reaching authority are summarized in Article 1 of the Compact, Findings and Declaration of Policy:

> . . . that the waters of Lake Tahoe and other resources of the Lake Tahoe region are threatened with deterioration or degeneration, which may endanger the beauty and economic productivity of the region.
>
> . . . by virtue of the special conditions and circumstances of the natural ecology, developmental pattern, population, distribution, and human needs in the Lake Tahoe region, the region is experiencing problems of resource use and deficiencies of environmental control.
>
> . . . there is a need to maintain an equilibrium between the region's natural endowment and its manmade environment, to preserve the scenic beauty and recreational opportunities of the region. . . .
>
> . . . it is imperative that there be established an areawide planning agency with power to adopt and enforce a regional plan of resource conservation and orderly development, to exercise effective environmental controls. . . .

Although it is not apparent from the Compact, the document is a tribute to the public concern over specific issues and broad environmental issues, as exposed in organized groups and widespread involvement.

The Compact Legislation

The structure of the Tahoe Regional Planning Agency is of major importance from a policy standpoint and is still a matter of discontent among some groups. The TRPA as created in the Compact is governed by a ten-person body, five each from California and Nevada, plus one nonvoting member representative of the United States appointed by the President. Six of the members are one each from the five counties and one from the City of South Lake Tahoe. Two are nonresidents of the Basin appointed by the respective governors, and two are the administrators of the California Resources Agency and of the Nevada Department of Conservation, or their respective designees.

The representation is six to four in favor of the local members of the counties and the City of South Lake Tahoe. Some critics make much of this factual point based on the general assumption that local jurisdictions are "developer oriented," which is not necessarily true. It is also often asserted that interests outside the general region are "conservation oriented"; this charge is unclear in meaning and also not necessarily true. The purpose of the mix was to give weight to responsible and knowledgeable people from the Tahoe area. Experience and records of voting have shown no consistency to bear out any of these concerns and feared prejudices.

An Advisory Planning Commission (APC) is appointed by the agency governing body. The total membership is a minimum of 17, including: six chief planning officers of the five counties and of South Lake Tahoe; two county directors of sanitation; two county health officers; two heads (or designees) of major state health and water agencies in California and Nevada; at least four lay members, each of whom is a resident of the Tahoe region; the Executive Officer of the agency serves as chairman. The APC is a working body that studies, investigates, and recommends plans, amendments, and other proposals to the governing body for its action.

The planning requirements of the Compact are specific: to cover the preparation of an interim plan and a continuing regional planning effort. Within 60 days after the formation of the agency, it was to recommend an interim plan and adopt one in 90 days. Within 15 months after the formation of the agency, the Advisory Planning Commission was to recommend a regional plan; in 18 months the agency was to adopt a plan and, with the commission, "continuously review and maintain" it.

In the wording of the Compacts the plan is to include the following correlated elements:

1. *A land use plan* for the integrated arrangement and general location and extent of, and the criteria and standards for, the uses of land, water, air, space and other natural resources within the region, including but not limited to, an indication or allocation of maximum population densities.
2. *A transportation plan* for the integrated development of a regional system of transportation, including but not limited to, freeways, parkways, highways, transportation facilities, transit routes, waterways, navigation and aviation aids and facilities, and appurtenant terminals and facilities for the movement of people and goods within the region.
3. *A conservation plan* for the preservation, development, utilization and management of the scenic and other natural resources within the basin, including but not limited to soils, shoreline and submerged lands, scenic corridors along transportation routes, open spaces, recreational and historical facilities.
4. *A recreation plan* for the development, utilization and management of the recreational resources of the region, including but not limited to, wilderness and forested lands, parks and parkways, riding and hiking trials, beaches and playgrounds, marinas and other recreational facilities.
5. *A public services and facilities plan* for the general location, scale and provision of public services and facilities, which, by the nature of their function, size, extent and other characteristics are necessary or appropriate for inclusion in the regional plan.

The Compact also states: "All provisions of the Tahoe regional general plan shall be enforced by the agency and by the states, counties and cities in the region." Clearly this was no minor assignment, particularly in light of the 18-month time frame and the limited financial resources.

The Compact is the source of authority for the agency, and it establishes standards through the development of implementing criteria, rules, regulations, and ordinances that are binding on the existing jurisdictions in the Basin. These local jurisdictions continue to carry on their normal functions in the Basin provided they at least meet the requirements set by the agency.

The items of control included in the Compact are: "water purity and clarity; subdivision; zoning; tree removal; solid waste disposal; sewage disposal; land fills, excavations, cuts and grading; piers; harbors; breakwaters or channels and other shoreline developments; waste disposal in shoreline areas; waste disposal from boats; mobile home parks; house relocation; outdoor advertising; flood plain protection; soil and sedimentation control; air pollution; and watershed protection."

The question of enforcement is similar to that for local jurisdictions in that the recourse for noncompliance is through courts of competent jurisdiction.

The agency does not have authority over federal units of government. Also, the Compact has a grandfather clause stating that any business or recreational establishment in the Basin that requires state license and was licensed as long ago as any part of the year preceding February 5, 1968, or is to be constructed on land that was so zoned or designated "in a finally adopted master plan" on the same date, is to be recognized "as a permitted and conforming use." Additionally, the agency has no powers of eminent domain, taxation, or legislation.

The Compact continues to excite debate among many political, citizen, and conservation groups. The continuing concern is centered around the issues of local versus regional representation and control and the limitations of the Compact in terms of the authority to implement objectives deemed desirable by certain conservation groups.

As town development and sophistication grows within the Tahoe Basin, the possibility of additional incorporation of cities also increases. New cities will require additional representation on the Agency Governing Body, thus furthering the balance in favor of local jurisdictions.

The permanent population of the Lake Tahoe Basin is approximately 28, 000, but the peak summer day population is approximately 135,000. These figures are also used in favor of greater regional or "outside" representation.

The Plan for Lake Tahoe

The interim plan was put together from existing material to meet the requirements in the Compact. Intended only to serve as an operating base, the interim plan was adopted on June 19, 1970. The planning program was hindered by a lack of funds, since both El Dorado and Placer counties withheld funds pending law suits against the agency. The suits challenged the constitutionality of the former California Regional Planning Agency and of the TRPA under the Compact. None of the suits was sustained, but the planning effort was hampered.

In May 1971 the agency staff released its first endeavors at a regional plan to the APC. The document might more properly have been called an environmental analysis of the Basin, for it did not contain the General Plan elements mandated by the Compact: a land use plan, a transportation plan, a conservation plan, a recreation plan, and a public service and facilities plan. The APC rejected the staff proposal as prepared under the direction of then Executive Officer Mr. J. K. Smith, and created a subcommittee comprised of the five County Planning Directors and the Planning Director for the City of South Lake Tahoe. Chairman Richard Heikka (who later became the new executive officer) was directed to prepare a plan within 60 days for the September 22 deadline of the 18 months specified in the Compact.

"The Plan for Lake Tahoe" was prepared by the subcommittee and consultants and adopted by the Tahoe Regional Planning Agency on December 22, 1971. The plan sets some quite precise limits to further growth through land use designations: General Forest; Recreation; Rural Estates; Conservation Reserve; Low Density, Medium Density, and High Density Residential, General, and Tourist Commercial and Public Services. These uses are further constrained by the "land capability map" prepared by the Tahoe Basin Planning Team of the U.S. Forest Service, which constitutes a part of the plan. The concept of implementing land capability constraints through limitations on impermeable surface coverage is a very progressive endeavor toward the preservation of environmental quality. The actual implementation of the plan was, and is, effectuated by four ordinances that were adopted soon after the adoption of the plan itself: a land use ordinance, a subdivision ordinance, a grading ordinance and a shoreline ordinance.

The General Plan also addresses the mandated elements of Conservation, Recreation, Transportation, and Public Services.

The originality and effectiveness of the "Plan for Lake Tahoe" lies in its strong relationship to the environmental capability of the land and its unique means of implementing standards designed to ensure environmental protection.

The Compact calls for the implementation of the general plan and plan elements by ordinances. This is a departure from the traditional concept of a long-range plan which is primarily goal oriented and is achieved (or not achieved) on an immediate basis by zoning maps and ordinances. Since there is no provision for zoning maps in the Compact, the long-range or ultimate plan for Tahoe automatically becomes the immediate or short-range plan also. More recent legislation in California requiring substantial compliance between zoning and general plans of cities and counties has brought local jurisdictions to face a similar dilemma.

The Plan has been variously described as being for 1980, 2000, and the ultimate development of the Basin. Since the Compact implies a definition of ultimate levels of development, the plan is really an attempt to deal with ultimate population levels and urban form. This is not to say that the plan cannot or will not change, but that any such change should be made toward that same concept. In the first year since its adoption, the plan has not been amended.

On what basis can one establish the limits of urban development or population levels when the environmental impact of development is determined as much by how as by what or where. In Tahoe this has been achieved by the two levels (or overlays) of control—the land use map and the land capability map.

The correlation between urban land uses (areas where major development may be permitted) and a high environmental capability of the land is extraordinarily accurate. The exceptions are primarily areas that under the terms of the Compact, qualify for "grandfather" recognition by virtue of the existing development or the previous granting of permission to develop. Thus the more intensive levels of development are generally constrained to those areas which are likely to be most tolerant.

The question of how a development is executed in relation to environmental protection is addressed by performance standards in relation to the varying levels of land capability. The ordinances dealing with these standards are primarily the land use and grading ordinances, and in the case of shoreline areas, the shoreline ordinance. Remembering that one of the major motivating forces behind the creation of the TRPA was the deteriorating quality of the water of Lake Tahoe, it is not surprising to find that the key performance standard—namely, limitations on impervious surface coverage—is directly related to water quality. The greatest delicacy of the whole Tahoe environment lies in the shallow soils formed from decomposed granite, highly erodible and extremely slow to revegetate. Large areas of impervious surface result in the channeling of runoff and the subsequent aggravation of all these sensitivities. Similar constraints on grading, including seasonal limitations and revegation requirements, serve the same ends.

The limitations on impervious surface coverage are such that in many instances the designated density limits of the land use plan cannot be achieved. Thus through the dual application of land use and land capability constraints, a rational approach to the calculation of population or the holding capacity of the Basin has been pursued. While the question of desirable or tolerable population densities continues to be debated and alternative levels to the 260,000 now projected are offered, none are computed from a more rational basis.

What Have We Learned from Tahoe?

With the exception of a detailed transportation plan, which was to have begun in 1973, the majority of the mandates of the Compact have been fulfilled or are well on the way to being fulfilled. Thus the question, What have we learned from Tahoe? can be examined in two ways: the inadequacies of the Compact in achieving the goals desired for Lake Tahoe by a majority of the people; and the learning experience of executing the legislative mandates that may act as a guide to other regional management endeavors.

The environmentalists continue to air their criticism and dissatisfaction with the TRPA and request the pursuit of a harder line on environmental issues. What these critics are really saying is that they did not like the mix of values represented by the political institutions of local jurisdictions and they do not like the mix of values promoted by the present regional system. One must guard against the assumption that these critics represent the majority viewpoint.

Undoubtedly, however, the plan and its implementation have been restrained by the limits of legislative authority. These major constraints can be summarized as follows:

- The TRPA has little or no control over its budget and for the tasks assigned to it, it is poorly funded.
- The TRPA has no taxing authority; thus it is deprived of a fundamental tool of planning—namely, real property taxation. Some of the products of this constraint have yielded severe economic impacts, resulting from the adoption of the plan on special service districts.
- The TRPA has no authority to own (buy, condemn, or receive) property. Without this power, the agency has been almost entirely dependent on the police power for its implementation, and $170 million worth of claims against the agency for inverse condemnation are witness to the limit to which this authority has been utilized. The agency is confident that these claims are ill founded, but the ability to implement agency conservation and recreation goals would be greatly enhanced by the acquisition of broader authority.
- The exclusion of the gaming industry from the authority of the TRPA has deprived the agency of control over the major growth-inducing element in the Basin.
- The existence of a plethora of special districts, including 106 water districts alone, with their various assessment practices tends to produce a motivation for development and wasteful duplication of public investments. The TRPA could benefit the Basin if it had authority over special districts, their formation and amalgamation.

As for what we may learn from a political and institutional standpoint that may contribute to the evaluation of the regional environmental management concept, the first lesson is that there is no panacea or pat solution. Some principles can be extracted:

- Whose or what interest should be served by regional management? Tahoe has strong local, regional, and national values and interests, and there are adequate supporters for all three factions. Those interests must be evaluated, and their balancing and blending is difficult. It is hard to acknowledge that there is no formula for the best solution.
- Wherever there are conflicting interests and conflicting and multiple demands on the use of land, no decision will satisfy everyone. There is a clear need to reach a tolerable level of acceptance and the success of the TRPA is largely founded in the ability of its present staff and the governing body to evaluate that level without forsaking its responsibilities.
- It is much easier to advocate, criticize, demand, or become cynical than it is to face the responsibility of cost, judgment, and decision. The advocates at Tahoe have been influential through the media, but the success of the planning effort is a credit to the decision makers.
- Of the two plans presented by successive staff administrations of the TRPA, one was successful and one was not. The heavy dependence on and investment in academicism and professionalism to solve a planning problem that clearly has a strongly political aspect and a well-documented political history proved to be ineffective. The plan that was approved in December 1971 was communicative, direct, and to the point of the agency's charge.
- Communication by and between the planners and the plannees who are affected is a critical prerequisite of plan adoption. The failure and success of the two administrations is directly related to their communicative abilities and the handling of the agency/citizen interface.

With these specific lessons must go the broader recognition of the impact of the TRPA and the gradual salvation or improvement of environmental quality. The path has not been a simple one and there is still much to be achieved, but the progress that has been made is now documented and supports the fundamental arguments for a regional environmental system.

To effectively deal with public issues such as land use, water quality, and environmental degradation, the political or administrative system must make certain kinds of decisions. The issues at Tahoe demand a holistic approach that is cognizant of the myriad interrelationships of the natural systems of the Basin. This kind of decision making was impossible under the previous array of functionally and areally fragmented boundaries of the local jurisdictions. The TRPA has provided the mechanism to deal with problems on an areawide basis and permits uniformity in the way issues are defined and resolved, and solutions enforced. Without this mechanism it would be impossible to realistically address the questions of population capacity, water quality preservation, and shoreline protection, for the failure of any one jurisdiction to uphold the desired standards substantially prejudices the success of other efforts.

Regional Functions

L. RUSSELL FREEMAN

At the national level, we tend to deal with fairly narrowly defined problem areas, such as transportation, environment, health, and welfare. Programs within these areas are even more narrowly defined. Thus we have a highway program, a sewer program, a social security program, and so forth.

It becomes possible to realistically think in general terms only at the local level (e.g., the city or the county). For example, the community master plan describes how physical features such as highways, sewers, houses, and businesses fit together in the community, and how it is proposed that new features will be added. The city budget represents a very real setting of priorities in terms of the day-to-day delivery of services.

The Regional Link

One function of regional government is to provide a transition between these two very different ways of approaching things. This function is carried out at three levels of regional jurisdiction (here using "region" loosely): the multi-community region, the state, and the multistate "federal region."

A number of things must occur at one or more such regional levels:

1. Someone must ensure that the various pieces of the system fit together within the community.
2. Someone must see that program and physical features of nearby communities are compatible.

Mr. Freeman was Director of the Office of Congressional and Intergovernmental Relations for EPA, Region IX. Currently he is Deputy Administrator for the Region. The views he expresses are his own; they do not represent EPA policy.

3. Someone must ensure that programs or features conform to national and state policy.

4. Someone must also be responsible for making cost-effective decisions.

5. Someone must allocate limited resources among competing communities.

6. Finally, in the creative sense, the system should provide a mechanism for creating the ability to respond to social problems.

One may question inclusion of the first item in this list, since it seems to describe the job of administration. Perhaps a few words of explanation are in order. The categorical approach at the federal level has tended to force similar categorization at the local level. Each federal bureaucrat wants a counterpart to deal with, a program to take credit for, and a plan to promote. In such a situation, actions are often dictated by circumstances (e.g., the availability of federal funds) other than local needs/or priorities. In extreme cases, projects can proceed in direct contravention of local plans and priorities. This multiple "program" influence often results in other conflicts as well. We have all heard of those cases of dams and highways, or parks and sewage treatment plants, or other features, competing for the same piece of real estate.

This type of situation has created the need for a regional overview of the local community planning process. This overview and many of the other functions listed have been assumed by the regional federal official because of his control over the available funds. The president's move to return control of funds to state and local officials thus presents an opportunity for decision making to return to the state, the local region, and to local officials. For example, revenue sharing permits greater flexibility in determining how problems will be solved or how services will be provided. Instead of choosing between alternatives such as building a highway versus losing highway funds, local officials will now be able to weigh more comparable alternatives such as building a highway versus buying buses with available federal funds.

Eventually, one would hope that local officials would move on to considering such broad options as providing employment opportunities near dwelling places versus providing effective commuter systems. Such a move to greater independence in decision making is possible under the president's "new federalism" program if state and local governments seize this opportunity to assume a truly autonomous role in the "system."

Of course, the federal government does have a legitimate continuing role in the system, even under new federalism. With the move away from control over federal funds, a new mechanism will emerge to provide for this federal role. The environmental program clearly illustrates one possibility—the establishment of a federal regulatory system. Through this mechanism, federal policy is established and state–federal implementing regulations are developed.

However, the action role is left to local government and private enterprise. It should be noted that the private citizen is given a strong role in this system through the public hearing requirements and citizen legal action provisions.

Intergovernmental Structure Effects

The effect of this overall approach should be a downward shift in the level at which the six regional functions are performed. The first and fourth functions listed may become local government processes entirely, with only limited review at the lowest regional level. The second, and perhaps even the third functions listed, may well be vested with the local region or perhaps the state. Ideally, the federal region will fall back to an overview role based on the need to conform to federal policy and priorities and the need to create new tools to deal with new problems.

Within this concept, it should be possible to greatly simplify the process by which local, state, and federal funds flow into the system for delivery of services within a community. It should be possible, for example, for the community to develop a land use master plan setting forth the layout of facilities and the zoning within its jurisdiction. This would become one basic document that passed through the regional, state, and federal systems for clearance with regard to framework planning and policy conformance. In addition, a community should be able to develop a single annual operating plan and budget. This plan would describe those actions intended to implement the community master plan, as well as the local program for delivery of services. It should be noted that development of such a plan corresponds roughly to the process that a city or county administration goes through in developing its annual budget. Again, this annual operating plan would be reviewed at the regional, state, and federal levels for conformity to policy and priorities, and for compatibility with other relevant plans. Assuming that these hurdles are cleared, this single annual work plan could also serve as the basis for receiving all state and federal implementing funds.

It remains to be seen whether such a simplified intergovernmental planning and budgeting system can ever be achieved. The outcome depends greatly on local initiative and on improved intergovernmental trust and cooperation. However, the opportunity has been presented.

In conclusion, it seems clear that since the move to a revenue-sharing system will greatly modify traditional intergovernmental relationships, we should begin to develop new relationships more appropriate to the new system.

Part II

Environmental Impact Statements

Patricia A. Bonner

In 1969 the National Environmental Policy Act (NEPA) was passed, forming the President's Council on Environmental Quality and presenting the nation with Section 102(2)(c), environmental impact statement requirements:

> The Congress authorizes and directs that, to the fullest extent possible: (1) the policies, regulations and public laws of the United States shall be interpreted and administered in accordance with the policies set forth in this Act, and (2) all agencies of the Federal Government shall.
>
> . . .
>
> (c) include in every recommendation or report on proposals for legislation and other major Federal actions significantly affecting the environment, a detailed statement by the responsible official on
>
> (i) the environmental impact of the proposed action,
>
> (ii) any adverse environmental effects which cannot be avoided should the proposal be implemented,
>
> (iii) alternatives to the proposed action,
>
> (iv) the relationships between local short-term uses of man's environment and the maintenance and enhancement of long-term productivity, and
>
> (v) any irreversible and irretrievable commitments of resources which would be involved in the proposed action should it be implemented.

States followed suit, passing their own environmental quality statements and acts requiring impact statements for publicly funded projects. Then came the task of establishing guidelines; training people to prepare statements and reports and to evaluate them; setting up procedures, budgeting, staffing, and so on.

Once governments had acclimated to, though perhaps not happily accepted all this, and once the process had become part of local government's infrastructure, another blow struck: the *Friends of Mammoth* vs *Mono County* decision in the California State Supreme Court.

All over the nation, governmental entities began examining the consequences of the decision, which required that impact statements be prepared for privately funded projects needing local governmental approvals. Which projects are exempt? How do you know whether a project will have adverse, irreversible environmental impacts that cannot be mitigated? How do you decide whether economic benefits exceed environmental disbenefits? What are the best administrative groups to manage the process?

Government, industry, and the public at large have only begun to wrestle with the intricacies of the impact statement process.

The National Environmental Policy Act

WILLIAM MATUSZESKI

The National Environmental Policy Act, signed into law as the first act of the president in this decade, accomplished three basic purposes. It set up the Council on Environmental Quality, which, unlike many agencies, survived Executive Office reorganization. It also expanded the missions of the existing agencies of the federal government to include within them a broader consideration of environmental values in their decision making. And it produced the action-forcing mechanism that is called for in Section 102(2)(c) and has resulted in the environmental impact statement.

I think that part of the interpretation of NEPA and what it means is related to some of the goings-on in Congress, particularly the Senate. There are two very active environmental committees in the Senate: the Public Works Committee, headed by Senator Edmund Muskie of Maine, and the Interior Committee headed by Senator Henry Jackson of Washington. To a large extent, the debates in the Senate have revolved around the personalities of the chairmen.

Mr. Muskie, through his staff and colleagues on the Public Works Committee, has generally favored specific national legislation that establishes very close guidelines, standards, and deadlines, and otherwise states exactly what is expected of officials at all levels of government over a long period of time, sometimes 10 to 15 years in the future. Obvious examples are the Clean Air Act Amendments of 1970, and the water bill passed in 1972 (PL 92-500).

Over in the Interior Committee a somewhat different approach is taking shape—one that is much more concerned with institution building and much less related to the specific requirements of a series of standards and dates to be

Mr. Matuszeski is with the Executive Office of the President, Council on Environmental Quality.

met to implement the law. The National Environmental Policy Act is clearly a law that passed the Interior Committee in the Senate with the broad institution-building concept in mind.

There are some confusing bits and pieces of the pie—in particular, how the National Environmental Policy Act applies to regulatory actions of the Environmental Protection Agency (under the Clean Air and Water acts).

In practice, the National Environmental Policy Act *has* affected decision making in government greatly, probably because it has made decisions public. It has allowed the public and other agencies of government to learn what was being decided on. And where NEPA has worked, it has made those decisions and those deliberations at a time when countervailing considerations can be brought in, perhaps resulting in a better project, or no project at all. If we look at NEPA in terms of an act primarily designed to bring in many people's differing points of view to improve decision making, we very quickly realize that NEPA works best when applied at the local level, when used as a decision tool by federal agencies and state and local officials.

The Questions of Interpretation

It is important to look at a few of the current interpretations of the requirements of Section 102. Everyone asks, What is a major federal action with significant impact on the environment? and How do you decide what they are? The Council on Environmental Quality guidelines make clear that the types of action include direct federal activities, such as a Corps of Engineers project, supported federal activities, (e.g., a grant to a locality to build a sewer), and federally approved activities (a license, a permit, for use of public lands). The greater questions become, What is significant? and How do you determine what is significant?

We are trying to come to some reasonable determinations that can be applied throughout the entire series of program actions, resulting in impact statements that are sufficient to cover the major activities of what occurs on a day-to-day basis, but not so many statements that they clog the system and overlap one another. Probably the best measures we have are some measure of mass—How big is the thing? What are we really talking about, in all its ramifications? How much money is involved? How much land?—Is the project indeed changing things? Is it inducing other changes? Will it result in different things happening now and down the line? For example, a major project might affect a rural area, whereas the sample project in an urban area, where a great deal of activity is already occurring, might not have the same significance.

The next question is, To whom does NEPA apply? It applies to federal agencies quite clearly. It applies to regulatory agencies of the federal government, as well. Perhaps the furthest that we have gone in that direction is a case that involved the National Capital Planning Commission (NCPC). As the Washington, D.C., planning agency, the NCPC operates as a federal agency when there is federal involvement (building of a federal building, the expansion of a park). It acts as the local planning agency when it must deal with other activities—approving changes in the comprehensive plan, for example. According to a recent court decision, it is not possible or conceivable for NCPC to shift roles. Thus it would be possible to conclude that if there is federal involvement, no matter what the basis is for involvement, the situation is covered by NEPA.

Next we must ask, How does NEPA apply to state actions? And here we have some guidance from the highway cases in the State of Texas, where the federal government was supporting the construction of interstate highway. Certain segments of it were clearly being funded by the federal government and went through an impact statement review.* A center section, however, was controversial. Texas determined that to avoid the NEPA requirements, it would fund this portion with state monies only. However, the court saw that the state contemplated tying in with the federal highway trust fund to finance the construction of this segment of the highway in the future. Therefore, it ruled that the act did apply to the state's action.

On the other side of this issue, an airport extension proposed in Boston was found not to be bound by NEPA because Massachusetts had not gone to the federal government for approval and funding. It is not clear whether the decision will stand over time, however, particularly since the same court—First Circuit, Court of Appeals in Boston—in February 1973 made a ruling that is more far reaching. The *Silver V. Romney* case goes beyond state action and examines private action. For the first time a federal court said something that has been implicit in the National Environmental Policy Act: private action in league with federal activity (in this case, federal mortgage insurance for a private developer's housing project) can be enjoined under NEPA because of the involvement of the federal agency in the project.

To *what* does NEPA apply? Clearly, to the natural environment. No one had any problems with that, and the manmade environment was early brought in. In a very good case in Virginia, *Eli vs. Veldey,* the court decided that putting a prison in a certain area might irreparably harm the historic nature of the neighborhood, which was a rural area with a great many antebellum mansions. The finding was that an impact statement was required.

The "work environment" is entirely unlitigated: it is a new and emerging area. Does the work environment require impact statements. Does the setting of standards for exposure to chemicals by workers by the Department of Labor require impact statements? The first case will be the reentry case of migrant workers into fields sprayed with certain pesticides.

Finally, we have the social environment. Clearly the National Environ-

* Editors' note: Environmental Impact Statements, as required by NEPA, are discussed in subsequent chapters of Part II.

mental Policy Act covers the social environment. The most recent case of interest is *Hanley* vs *Kliendintz* in the Second Circuit Court of Appeals of New York, where it was found that the agency had to consider whether an impact statement was needed for a federal prison facility, and that among the considerations should be the effect on the neighborhood.

One other point bears mention here. Most people expected that when the National Environmental Policy Act was passed, it would deal primarily with the natural environment; they thought that the Interior Department activities clearly should be covered, as well as those of the Agriculture Department and the Corps of Engineers. But when we examine the litigation pattern, we find that the vast majority of active cases are related to urban areas. The most litigated agency is the Department of Transportation, which is to be expected because of the high volume of impact statements produced. Most of those cases involve urban highways. The second most litigated agency is the Corps of Engineers. However, the interesting ones are three and four—Housing and Urban Development and the Environmental Protection Agency. Thus the National Environmental Policy Act is having broad implications and is being applied by citizens within urban and urbanizing areas.

The Future for NEPA Processes

To *what* extent does the NEPA apply? Here we have some problems. Most impact statements cover primary impact quite well. Much more work needs to be done in the area of secondary impacts. We are now investigating ways of evaluating the growth-inducing impact of interceptor sewers and highways. This area will probably receive more and more emphasis. We are learning that for many public facility projects, the primary impacts—mainly, those that are directly a result of the construction activity—are far less important than the growth-inducing, or secondary impacts. Some people have raised a tertiary level. They say that the impact statement should look at growth in general and should attempt to determine how the project is going to affect the overall regional population, not just its distribution but the population itself and the overall ecosystem of the area.

Integrating the impact statement process with the actual decisions on the project is probably our biggest problem. Some of the approaches we have developed include the "negative declaration" and the "notice of intent." The negative declaration is simply a statement for the record that the impact of the project has been examined and it has been determined that an impact statement is not needed. Sometimes negative determinations have become quite detailed; it is not unusual to see one 15 or 20 pages long. A determination may conclude "no impact statement needed," but if it was 20 pages long, you are a little bit suspicious. Why did it take 20 pages?

The notice of intent, on the other hand, is a very useful tool because it says basically, "We are going to do an impact statement. Let us know if you have got any information to prepare it." Timing is the key. It is best to get a notice early enough to allow yourself to affect the choice of project and choice of design of the project, and yet late enough that you know what you are doing.

We are faced now at the federal level with two sets of problems. Some agencies (the Corps of Engineers and the Department of Transportation) are producing too many statements that are not genuinely comprehensive and analytical, particularly in terms of the secondary impacts. In other agencies, the production level is simply not high enough, and everybody says "Don't talk about production with NEPA; don't say anything about how many statements are done." But if after three years we find that the Environmental Protection Agency has produced 13 statements for the whole country in one year, we must start talking production. EPA and HUD are two agencies which simply must do more work.

In terms of applying NEPA to the regional development of an area, I think that there are four areas for consideration. First, it is important that the regional agencies have a way of identifying the federal actions subject to NEPA. There are three good ways of doing that now, one is the A-95 Review (all programs reviewed by a local agency designated by federal government's Office of Management and Budget—comprehensive evaluation of proposed program prior to approval of any federal grant) which comes before an impact statement is prepared; another is the notice of intent which an agency should issue when it decides to do an impact statement and requests help; and the third is the draft impact statement iself. At those three levels there should be input. There should be people working to draft a statement that will be meaningful for people to review, and others to collect the comments and produce a final draft that will stand up to criticism. Second, we need to analyze individual actions and consider the broader implications of actions. Not enough regional agencies and planning agencies provide comments that reveal thorough examination of the full affects of the projects being reviewed. This is particularly true of the Department of Highways; the job is not being done in a good many parts of the country. Third, more could be done to use federal agency planning mechanisms. The Corps of Engineers has regional planning. HUD has its 701 urban development grants program under the Housing Act of 1954. EPA does comprehensive regional planning. Transportation has its highway planning. There is relatively little coordination among these planning activities at the federal level, but at the regional level there is an opportunity to bring a program all together. And the impact statement can bring it together because everyone who is doing regional planning has the opportunity to comment on everyone else's work. Finally, there may be some opportunities for using the impact statement to look at multiple-agency impacts on a region. For example, we might look at all the activities that all the federal agencies are planning the next several years in a given region and estimate what the future conflicts could be. To a large extent, this is an approach we have taken with the northeastern panhandle of West Virginia, where they are heavy water development pressures, water resource development pressures, vacation home pressures, a national forest, and some Federal Power Commission projects. There may be a chance to use impact statements in considering these multi-agency impacts.

Legal Aspects of Impact Statements

NICHOLAS C. YOST

This chapter reviews CEQA (California's Environmental Quality Act) and presents some of the legal developments which I think will take place under both the California and the national acts.

In the time since the National Environmental Policy Act was signed by the president on January 1, of 1970, 11 states plus Puerto Rico have adopted their own "little NEPA's." That is a lot of laws in a short while, and there is every reason to believe that other states are also going to be adopting them. Thus the experience of California, having been the first state to adopt a "little NEPA" will be useful to planners in other states.

Largely because of the California Supreme Court, I think, our NEPA, CEQA, has come to be the most pervasive in the nation. And in a sense, since it finds application in all local land use decisions, it is more pervasive than NEPA.

California Environmental Quality Act

Also enacted in 1970, the California act was clearly patterned on the National Environmental Policy Act. However, there were a few differences. CEQA used the "project" instead of "major action" as the key term designating the thing that occasions an environmental impact statement. And this is what gave rise to the debate over what the act is supposed to cover. Basically, the State Resources Agency and the State Office of Planning and Research (which was charged with promulgating guidelines) had said that the act applied to Public

Mr. Yost, a Deputy Attorney General for the State of California, has the chief responsibility of making CEQA work.

Works projects (i.e., the "directly undertaken" type of project—building a highway, building a dam, etc.). It was the view of our office, the Attorney General's office, that it also included the public permission of private activity, where a permit or a lease or funding, but some form of governmental connection with what was happening on the private level.

We had three reasons for taking the position just outlined: (1) that is what the author said he meant, (2) there were policy statements at the beginning of the California Environmental Quality Act that could be implemented only if it were interpreted this way, and (3) in looking at the legislative intent, and particularly in defining the term "project," we referred to the guidelines that had been adopted by (CEQ) the Council on Environmental Quality for the Federal act, which used the term "project" and used it in the sense that we used it. A public dispute was stirred up, and meanwhile most local governments were not applying the state act to publicly permitted but privately undertaken activity.

The Friends of Mammoth Case

Environmental lawyers started looking around for a case. For various reasons, largely because of uncongested court calendars, Mono, a remote, and pastoral county on the eastern side of the Sierras, saw the first case. A group of people who had small, second-home cabins in a fairly fragile area were opposed to the construction of a large condominium—actually two condominiums, which were going to be the first of half a dozen seven-story buildings stretching over three-quarters of a mile—in their mountain area. They went to court; they lost in the trial court; they lost in the court of appeal; and their rather ingenious attorney, who seemed to know only that he wanted to stop the project, filed with the California Supreme Court something called "A Petition for Appeal, or in the alternative for hearing, or in the alternative for supersedeas or in the alternative for mandate, or in the alternative for. . . ." The petition title went on like that, naming everything the attorney could think of, and the court granted it. Thus the California Supreme Court became the first appellate court to consider the direct issue of whether CEQA applied to private activity.

Then the court became briefed by everybody. We went in as a friend of the court, the Sierra Club went in as a friend of the court, and then having been stayed by the court, the matter was argued. Finally in September 1972, the decision came down, holding that the California act did apply to the public permission of private activity.

In the aftermath there was considerable confusion. Building freezes were imposed on a large number of localities, and there were comments to the effect that this was going to bring economic disaster to California. In some cases the building permit freezes were lifted; in Los Angeles County they were lifted within two hours after they had been imposed. But the headlines told of their imposition, not of their being lifted. There came to be widespread hysteria over what in fact was meant by the act. And the act, like the federal act, contains the limiting phrase "significant effect on the environment." In other words, every little project that can be imagined—somebody adding a room, for example—does not occasion an Environmental Impact Statement. You have the overall requirement to be sure that matters that would make a difference in the environment in which people live are considered, and not the others.

The Supreme Court had perhaps contributed a bit to the confusion when it chose "nontrivial", as a synonym for "significant." Many people grasped the term "nontrivial" and said, Well if we have to do something for everything other than trivial, CEQA's effects are going to be widespread. At this point some political leaders within the state (in what the Attorney General referred to as manufactured hysteria) began to call for the repeal of the Environmental Impact Statement requirement in general, or the repeal of the act as it applied to private activity more specifically. Here I think that we were trying to see what legitimate problems people did face in having to cope with new processes, but at the same time preserving a method, the Environmental Impact Statement, of ensuring environmental preservation. And here we had the effective combination of Assemblyman John Knox, who chairs the Local Government Committee, and our office. Mr. Knox had a bill which he was carrying at the request of our office (which made it bipartisan, since the Attorney General and the assemblyman are of different parties). Together we tried to determine everybody's interest in the act and to decide how it could be channeled constructively to obtain legislation that would deal with the problems without repealing the act.

The CEQA/Mammoth Compromise

At this point, the court refused to reconsider the *Friends of Mammoth vs Mono County* case, and it deleted the reference to "nontrivial," which alleviated the fears of many. The legislature returned right after election, and everybody started giving as much input as possible to Assembly Bill 889; at one point the two principal opponents were the California Chamber of Commerce and the Sierra Club, which are perhaps improbable allies. However, as the bill ultimately passed, it had the support, or at least the nonopposition of every significant group in California, including business, labor, local government, and environmentalists. In other words, there was a bill that satisfied nobody but in which everybody felt there was a bit more for him than there was for the other people, and for that reason all supported it.

I will now briefly outline the elements of this great compromise in California law. For business and labor there was 120-day moratorium that applied to private, not public, activity. It validated past actions unless they were under judicial challenge, it provided for state guidelines to be adopted within 60 days, and local guidelines within 60 days thereafter (i.e., after the first 60 days), and it provided a statute of limitations. Thus after a decision is made there is a period during which suit can be brought, but after that period is over, the developer is free from the possibility of suits.

The local government was glad to see the 120-day moratorium, which they wanted perhaps more than anyone else. Their position seemed to be, "We'll be willing to live with this, and it may even be a good idea, but we don't have staff, we don't have any procedures; we must have a little time to develop them." Local government also received the certainty that would be provided by the guidelines, it got specific statutory authority to charge the private applicant for a permit and the cost of preparing the Environmental Impact Statement. In other words, the responsibility still rests with the local government, but they can pass on the cost, which becomes internalized to the project and is not borne by the taxpayer. The statute of limitations was also well received by local government.

The environmentalists applauded the preservation of the Environmental Quality Act, and the preservation of the *Friends of Mammoth* vs *Mono County* decision. For them there is another matter of extreme importance. The California act contained an exception by virtue of which any local government with a so-called "conservation element" did not have to prepare an Environmental Impact Statement but instead could make a finding of conformity to a conservation element. A conservation element was a very wishy-washy thing, not likely to provide significant protection for the environment. Also, since every local government had to have a conservation element by the middle of 1973, this would mean that in effect, *Friends of Mammoth* would automatically be repealed by the local governments one by one, as they adopted the exception. The conservation element exception was simply taken out of the act.

Finally, the "growth-inducing impact" was put in as a requirement for both state and local governments to consider as a specific element. "Growth-inducing impact" was implicit within the national act, and it is implicit within the other state acts patterned on the national act. The California act now has listed it as one specific item to be considered in our equivalent of a 102 statement.

The Future

Where are we now? We have had a moratorium, state guidelines have appeared, and the guidelines of other state and local agencies are being or have been developed. In going into the trends and case law, it is convenient to start with guidelines. There was considerable dispute at the time the Resources

Agency held public hearings throughout the State to hear comments on what are now the state guidelines, especially since the Resources Agency itself questioned the legal adequacy of the proposals. Business groups, environmental groups, and our office commented on the guidelines, which have been adopted, and they are surely not satisfactory to everybody. The California Supreme Court, at just that time in its second case involving the California Environmental Quality Act (*Desert Environment Conservation Association* vs *Public Utilities Commission*) ruled that it would be premature for the court to step in and rewrite the regulations for one state agency, the Public Utilities Commission. In declining, however, the court inserted a couple of sentences that were perhaps gratuitous, but at any rate gave some idea of what the court would be looking for. The court indicated that its decision was not to be construed to mean that regulations under CEQA would be immune from judicial review once adopted. An administrative regulation authorized by statute is invalid unless it is consistent, not in conflict with the statute, and reasonably necessary to effectuate the purpose of the statute. A regulation that fails to satisfy this standard is properly subject to judicial challenge. Unquestionably, then, state and local guidelines and the judicial challenges to them are going to be preoccupying California environmental lawyers for some time.

The second major issue appears to be the ultimate issue under both the federal act and the state act. You bypass the simple questions the courts have already dealt with. If there is no Environmental Impact Statement, you point out that the courts have said that there *must be* an Environmental Impact Statement. In the so-called second generation of cases there is an Environmental Impact Statement but it is inadequate, and the courts have said that an adequate statement must be prepared. And we are now entering the third generation of cases: according to an adequate report, properly prepared and honestly done, there will be a highly adverse environmental impact, but the decision maker decides to go ahead anyway. What happens then? This is likely to be the great area of litigation that will occupy lawyers in the field for some time to come. Obviously, courts cannot step into all the thousands of decisions that are made on local, state, and national levels. There are just too many decisions; nor is it appropriate for them to intrude. Lawyers devise phrases like "clear abuse of discretion" to refer to those cases in which an area has been terribly misused. And in those cases the court is going to come in.

The California Supreme Court in *Friends of Mammoth* gave a hint of where it was going in its now famous footnote 8, which has gone further perhaps than any federal court: "Obviously, if the adverse consequences to the environment can be mitigated, or if feasible alternatives are available, the proposed activity, such as the issuance of a permit, should not be approved." These rulings are what lawyers call "dicta"; it is not necessary to the decision in the case; however, it indicates where the court is going. The Federal Circuit

Court for the 8th Circuit has taken similar positions, saying that NEPA is intended to effectuate substantive change in decision making.

The court in *Friends of Mammoth* continued: "The unequivocal intent of NEPA is to require agencies to consider and give effect to the environmental goals set forth in the Act, not just to file detailed environmental studies which will fill governmental archives." I think that a balancing is occurring. We have the practicability on one hand, as well as the wisdom of becoming involved in every decision. On the other hand, the act, if it is to mean something, must be enforceable. It is no use to prepare an Environmental Impact Statement if that statement implies no obligation to act in an environmentally sound manner.

Finally, I think that the court's perception of its own role is very important in this area. Justice Mosk, in *Friends of Mammoth*, paraphrasing Judge Skelley Wright in the Federal court in *Calvert Cliffs vs the Atomic Energy Commission* said "The duty of the judiciary . . . is to assure that important environmental purposes, heralded in legislative halls, are not lost or misdirected in the vast hallways of administrative bureaucracy." Whatever directions NEPA and CEQA take, I am sure that they are going to be interesting, and many of us are going to be involved in the coming years.

Impact Reports and Local Decision Making

DAVID C. NIELSEN

Environmental Impact Reports in California

The California Environmental Quality Act of 1970 became law late in the year. With only minor exceptions, the act was patterned after Section 102(2)(c) of the National Environmental Policy Act.

For local government, Section 21151 was most important. It read as follows:

> The legislative bodies of all cities and counties which have an officially adopted conservation element of a general plan shall make a finding that any project they intend to carry out, which may have a significant effect on the environment, is in accord with the conservation element of the general plan. All other local governmental agencies shall make an environmental impact report on any project they intend to carry out which may have a significant effect on the environment and shall submit it to the appropriate local planning agency as part of the report required by Section 65402 of the Government Code.[1]

At the time, most, if not all, local decision makers defined the word "project" narrowly to mean an activity directly undertaken by government, such as the construction of a freeway or an aqueduct.

The Supreme Court of the State of California in September 1972 ruled against this interpretation of the act. In its decision in *Friends of Mammoth* vs *Mono County* the court ruled that an environmental impact report must also be prepared prior to the issuance of any permit or other similar entitlement

Mr. Nielsen was Acting Director of the Office of Environmental Management of the County of San Diego; he now is with the County's Human Resources Agency.

that could have a significant impact on the environment. In other words, the environmental consequences of most governmental decisions must now be subjected to analysis and public scrutiny prior to approval. Of most significance to local government, environmental analysis and public review must precede major land use decisions, such as the following:

1. The adoption of general plans and plan elements.
2. Zoning ordinance amendments.
3. Approval of subdivision maps.
4. The granting of variances.

Subsequently, the legislature in Assembly Bill 889 amended Section 21151 of the Environmental Quality Act to require environmental impact reports even in those jurisdictions having an adopted conservation element.

Responses by local decision makers and the general public to the California Environmental Quality Act and the *Friends of Mammoth* decision have been as intense as they have been varied. To some, the new requirements mark the beginning of a new era of more environmentally sensitive decisions by local officials. Others, believe that a threat has been posed to the State's economy. For local agencies, it represented a stern test of organizational and management skills in a climate of rather intense conflict.

In the short time that has passed since the Supreme Court decision, there has been little time to realistically assess some of the important issues surrounding the environmental impact report process. These include:

1. The utility of the impact report process as a regional decision-making tool.
2. The relation of the impact report requirement to the political process.
3. Problems of implementation and refinement.

What follows are random comments on certain aspects of the environmental impact report process and its implications at the local level. The analysis is based to a great extent on the author's experience with the County of San Diego's Integrated Regional Environmental Management (IREM) project. The project provided the county with the means necessary to comply with the requirements of the Environmental Quality Act earlier and more systematically than most other local jurisdictions. Hopefully, the work of the project staff will help to refine and place in perspective the potential of the environmental impact report process.

A Regional Decision-Making Tool?

To understand the impact of the Environmental Quality Act on regional decison making, two major features of the act must be known. First, it assigns responsibility for the preparation and review of impact reports to local governments (cities, counties, and special districts) and establishes no meaningful standards or criteria for region- and statewide use. Thus the quality and nature of reports varies dramatically from jurisdiction to jurisdiction.

The second point is that the environmental impact report process is not intended to be a regulatory device. Rather, it is an informational and procedural mechanism. Concerning this point, the state guidelines read as follows:

> *Informational Document.* An Environmental Impact Report [EIR] is an informational document which, when fully prepared in accordance with the CEQA and these Guidelines, will inform public decision makers and the general public of the environmental effects of projects they propose to carry out or approve. The EIR process is intended to enable public agencies to evaluate a project to determine whether it may have a significant effect on the environment, examine and institute methods of reducing adverse impacts, and consider alternatives to the project as proposed. These things must be done prior to approval or disapproval of the project. An EIR may not be used as an instrument to rationalize approval of a project, nor do indications of adverse impact, as enunciated in an EIR, require that a project be disapproved—public agencies retain existing authority to balance environmental objectives with economic and social objectives.[2]

Thus the findings of environmental impact reports are not binding on local decision makers, but rather are to be balanced against other goals and priorities.

In spite of these limitations, based on San Diego's experience to date, the process does hold some promise as a means of introducing regional concerns into the local decision-making structure.

In this regard, the state guidelines require that all agencies having special expertise or who may be affected by a local project be consulted during the development of the environmental impact report. By initiating such consultations, county staff has already been in a position to more effectively place projects in a regional context. At a minimum, decision makers have been supplied with new information about the effects of projects on the multitude of jurisdictions and interests in the San Diego region.

Second, the EIR process has led to the development of a better project monitoring capability within local government. In San Diego already, the Environmental Impact Analysis Section staff has developed an impressive information base relating to the location and nature of major development activities in the county. Staffs such as this have the potential to become key elements in the clarification and anticipation of regional growth and development trends. Effectively used, the information could be of considerable value to ongoing planning and program development processes.

Third, considerable new data have been generated in the process of conducting environmental analyses. Organized properly, this new data base could help in the development of meaningful standards and criteria for region-wide use.

CEQA and the Political Process

The majority of local governmental decisions affecting the environment are made based on criteria that are, to varying degrees, political. This is true because of the lack of clear consensus regarding land use priorities and goals. Environmental management at the local level is an imprecise tool as much for this reason as for the lack of sophisticated methods of data gathering and analysis. This being the case, it is of primary importance to assess the role of the impact report requirement in the political process.

Its apparent impact in this regard has been to clarify political issues by providing a vehicle for articulating environmental concerns. The impact report process has served to institutionalize communication between environmentally concerned citizens and local officials. Concerned citizens now have the right and an identifiable means to monitor and to a degree at least influence most major land use decisions. In this respect, the Environmental Quality Act has become the "environmental full disclosure law" for the people of California.

Implementing CEQA

San Diego County has experienced several problems in implementing the requirements of the Environmental Quality Act. Some of these will be solved over time as staff gains more experience with complex problems and procedures. Others are indicative of basic limitations in the act and the decision-making structure of state and local government. Taken together, they point to a full program of clarification and refinement of basic concepts and processes. Some of the most critical problems identified to date are as follows:

1. The lack of uniformity of local procedures.
2. The lack of criteria for evaluating environmental impacts.
3. Integration with the planning process.

Lack of Procedural Uniformity

Local procedures vary considerably from jurisdiction to jurisdiction. The result—general confusion on the part of the public—has tended to discredit the process and to irritate project sponsors already faced with an astonishing array of bureaucratic rules and procedures. It seems obvious that procedures must be standardized as quickly as possible.

The Criteria Problem

At present, no criteria exist for making even the threshold determination of environmental significance (i.e., at what point an EIR should be required). The County of San Diego has not developed quantifiable criteria for such determinations, although the problem is presently under study. Implicit criteria have of necessity been developed, and a degree of uniformity has emerged in staff determinations.

Attempts have been made by other jurisdictions to develop clear-cut, easily understood criteria. In most cases they have been reflective of project scale—for example, the number of dwelling units in a subdivision. Unfortunately, a single criterion based solely on project scale is not sensitive to project location and local policies and priorities. To reflect environmental issues meaningfully, adequate criteria will have to be developed which take into account all three factors.

Integrating the Environmental Impact Report and Planning Processes

The basic issue in the *Friends of Mammoth* decision was the application of the impact report process to major land use decisions. Since the decision was handed down, much staff effort has been devoted to integrating the impact report requirement with the land use planning process. That process is relatively complex and procedurally inflexible. Most of its procedures and time requirements are defined in state law and local ordinance. The addition of a totally new requirement has in some cases led to confusion. Much of the initial work by staff was aimed more at meeting legally mandated deadlines than at providing meaningful input into major land use decisions.

Now the environmental impact report process is assisting in decisions on project approval or denial. However, perhaps a more important step is yet to come—namely, the use of impact analyses to determine and set more environmentally sound conditions for project scale and design, and to contribute to traditional land use planning goals and priorities.

Conclusions

Recent events in California indicate that the environment is an issue weighing heavily on the minds of many. The Environmental Quality Act and the impact report requirement have responded to certain dimensions of the problem. Better information is now being provided to local decision makers. The public has a more effective forum for expressing its concerns.

However, the major issues have yet to be acted on in California. These include more effective and equitable land use controls, property tax reform, and local government consolidation and reform. Hopefully, the environmental impact report process will be able to contribute to these more fundamental issues by helping to clarify them and to add to our presently limited informational base.

Notes

1. George H. Murphy, Legislative Counsel, *Laws Relating to the Protection of Environmental Quality 1970*, State of California, Department of General Services, Documents Section, p. 14.
2. State of California, Office of the Secretary for Resources. *Guidelines for Implementation of the California Environmental Quality Act of 1970*, February 5, 1973, Section 15012, p. 3.

HUD's Implementation of NEPA

DALE T. JAMES

The Department of Housing and Urban Development (HUD) has always exhibited a certain environmental consciousness. To be sure, the focus was and essentially remains the man-built environment of urban regions. Within this general focus we have been particularly concerned with the immediate impact of the built environment on the quality of urban life. This focus has its basis in the goal of a "suitable living environment for every American family," which has been a key mandate of the department and its predecessor, the Housing and Home Finance Agency, since 1949.

I do not want this somewhat self-laudatory introduction to give the impression that the department is resting comfortably in the knowledge of the historic correctness of its philosophy and its procedure in environmental matters. Perhaps too often we hear the assertion that progressive new legislation, such as NEPA, is in fact quite consistent with an agency's policy and practices. HUD is a man-made institution, and man-made institutions have a way of being less than perfect. A great deal of work remains to be done before we can say that we are fully responsive to environmental factors in every case. This knowledge sustains a quite lively debate within the department and a persistent effort to find better ways of meeting our environmental responsibilities in the face of complex and often competing societal goals.

Comprehensive Planning Assistance Program

In HUD it was no accident that the responsibility for departmental implementation of the National Environmental Policy Act was assigned to the Office

Mr. James is the Environmental and Standards Officer for HUDs Region IX.

of the Assistant Secretary for Community Planning and Management. This is the office responsible for, among other things, the administration of the Comprehensive Planning Assistance Program. Early in the formulation of a departmental strategy to implement NEPA (1969) we recognized the intimate relationship between environmental protection and the comprehensive planning process. Many years of administering planning programs had convinced us of the inherent wisdom of the urban planning profession's historic concern with the interrelatedness of man's activities and the form and quality of his urban environment.

The Comprehensive Planning Assistance Program is particularly pertinent to the problems of land use control and environmental protection, since it provides an almost unique form of federal assistance to aid virtually all other levels of government to carry out comprehensive planning and to deal with the problems of land use control through an ongoing planning process. This program originated with Section 701 of the Housing Act of 1954. The rate of appropriations has increased over the years, and by July 1972 nearly a half-billion dollars wiil have been disbursed to a wide variety of planning agencies at all levels of government. Recently in the Comprehensive Planning Assistance Program, environmental considerations have been given increased emphasis, and they play a major part in grants under the program.

The program's environmental goals as defined in the regulations are to:

1. Improve and conserve the quality of the air, water and earth resources for the benefit of present and future generations in the planning and shaping of man-made environments;
2. Assure that environmental concern and awareness becomes an integral part of the comprehensive planning process, since comprehensive planning is a major means for accomplishing community development on a sound environmental basis; and
3. Achieve those goals set forth in the National Environmental Policy Act of 1969. . . .

The regulations further require that an environmental assessment be made to identify the salient man-made and natural environmental elements and to assess environmental factors which will:

1. Minimize or prevent undue damage, unwise use, or unwarranted preempting of natural resources and opportunities;
2. Recognize and make prudent allowance for major latent environmental dangers or risks (e.g., floods, mudslides, earthquakes, air and water pollution);
3. Foster the human benefits obtainable from use of the natural environment by wise use of the opportunities available (e.g., use of natural drainage systems for park and recreational areas);

4. Seek under the above policies and goals to:

(a). Avoid adverse environmental impacts on neighborhood or community areas through the planning and careful location and development of community facilities;

(b) Provide environmental amenities to all areas being planned for, and access to each amenity;

(c) Equalize the impact and burden of community change and development on living areas rather than concentrate them in areas where sites are cheap: and

5. Incorporate state environmental policies and standards, particularly those developed in response to federal law regarding protection of air and water quality and the control and abatement of noise.

The regulations also require an environmental assessment of the work that results from the Planning Assistance programs, particularly for developmental policies affecting land use development and arrangements, major community facilities, and utility and transportation systems. The assessment is identical with the type required under Section 102(2)(c) of the National Environmental Policy Act, commonly known as an Environmental Impact Statement, and is to be appended to the resulting plan and used in the public deliberations leading to adoption or approval of the plan. In this fashion, the environmental consequences of the proposed land use or other plan can be made known to the public and given due consideration before that plan becomes public policy and is completed.

We also have an active research and development program in the environment area, and we have cooperated with federal, state, and local agencies in pursuing innovative matters of mutual interest. For a number of years now we have worked with the Department of Transportation and agencies on the problems of urban and aircraft noise, producing a significant set of studies culminating in the promulgation of the department-wide policy setting noise standards for all HUD programs.

In California a number of significant projects have been initiated relating land use planning and environmental factors. One of the most significant of these is the Joint Bay Region Environmental Resources and Planning study. This regional-scale effort is currently under way through a federal–regional partnership represented by HUD, the U.S. Geological Survey, and the Association of Bay Area Governments. The partnership represents a convergence of the major federal producer of earth science environmental information, the federal department most concerned with urban regional land use planning, and a major user of environmental information, constituted as a regional agency responsible for comprehensive regional land use planning. The study is focused on the development of earth science environmental data and determining the most effective ways of introducing such data into the land use decision-making processes of local government. At the state level a somewhat parallel effort was initiated between HUD, the California State Office of Planning and Research, and the State Division of Mines and Geology. This project sought to identify major geologic problems statewide. Areas of future urban growth were determined, and the geologic problems were defined in terms of potential costs accruing to both present and future urban development. The findings are being used by the Office of Planning and Research in its work in land use policy formulation and by the Division of Mines and Geology as a basis for identifying data needs and establishing data collection priorities. At the county level, work under way in Sonoma County focuses on alternative patterns of urban growth and the application of cost-benefit methods, including natural environmental factors, to the urban land use planning process. In Los Angeles County workers are attempting to determine ways of sensitizing the urban land use planning process to earth science environmental factors and in particular using the events and experiences of the recent San Fernando earthquake. We are also involved in joint research with the National Oceanic and Atmospheric Administration of the Department of Commerce, the U.S. Geological Survey, and the National Bureau of Standards to reduce earthquake hazards through improved land use and building practices. This matter has been of considerable concern to our Federal Insurance Administration. We have been working with the EPA Office of Solid Waste Management on innovative technologies in waste handling and disposal, particularly with HUD's Operation Breakthrough. The development of total energy systems employing waste heat, the problems of mine surface subsidence, and continued involvement in the urban noise problems are other areas of continuing research and technology concern.

HUD's Response to NEPA

Federal action in response to environmental and pollution matters remained at a low level until 1969. Laws were passed and federal assistance and regulation of a sort were provided for water pollution, air pollution, and solid waste disposal. However, in comparison with the magnitude of the problems involved—such as the pollution of Lake Erie—the public efforts were extremely limited. Action lagged, partly by reason of the limited statutory authorities, partly because of limited funds, but primarily because of the very limited public support. In three years, all this has changed. Now there is lively concern with environmental matters. Congress is passing laws of more appropriate scope and power. Citizens and interest groups have challenged, on environmental grounds, many public actions, and the number of environmentally based court cases had increased dramatically.

HUD Organizational Arrangements

Office of the Deputy Undersecretary

In June 1970 HUD Secretary George Romney delegated to the Deputy Undersecretary the responsibilities for implementing Public Law 91-190, the National Environmental Policy Act of 1969, including internal HUD administration, relations with other departments, and relations with the Council on Environmental Quality. Mr. Romney asked each Assistant Secretary and the General Counsel to provide support in such matters relating to his respective program area or legislative functions.

Subsequently, the Deputy Undersecretary asked each Assistant Secretary and the General Counsel to establish points of contact for environmental matters; these designees became the Environmental Clearance Officers Group, which is the organizational means for providing staff support.

Office of the Assistant Secretary for Community Planning and Management

In October 1971 the Secretary signed a memorandum noting that the Office of the Deputy Undersecretary had completed establishment of initial policies and procedures, and he transferred those environmental functions to the Office of the Assistant Secretary for Community Planning and Management, adding this provision: "Within Assistant Secretary Jackson's Office, these activities will be administered by the Office of Community and Environmental Standards (formerly the Office of Community Goals and Standards), and technical support activities will be handled by the Environmental and Land Use Planning Division." Among the other environmental functions specifically assigned Community Planning and Management were those concerned with noise abatement and control.

Certain modifications at regional and area office levels were also intended; thus a vertically integrated system was ultimately established, utilizing Community Planning and Management counterparts wherever possible as the Environmental Clearance Officers and/or their supporting staffs.

Regional, Area, and Insuring Offices

At field level, lead responsibility for environmental functions is of course assigned to regional administrators and directors of area and insuring offices. However, specific Environmental Clearance Officer designations have been made which place the staff responsibility on the Assistant Regional Administrator for Community Planning and Management at the regional level, the Assistant Director for Planning and Relocation at the area level, and at the Insuring Office level through the land planning division which is under the direction of the Chief Underwriter.

This location of environmental responsibilities in combination with the pattern of a vertically integrated system reflects HUD's experience, which showed that our primary concerns in the environmental area are closely tied to areawide comprehensive and functional planning. In making environmental evaluations of HUD projects or reviews of other agencies' 102 statements, the principal aspects to be considered usually include consistency between project planning and other, broader area planning; impact on existing and planned land uses; and impacts on human activities that "significantly affect the quality of the human environment," to quote from Section 102.

We consider this effort by HUD to participate effectively in the federal action to protect and enhance the environment to be a missionary movement and an educational effort, not merely a routine review-and-approval process.

HUD Procedures: NEPA Compliance

The environmental activity required by NEPA which is most important to HUD, and I would assume most federal agencies, is the processing prescribed by Section 102(2)(c). As mentioned earlier, the section requires the federal agency to make a thorough environmental analysis and evaluation of the proposed project (or other undertaking), prepare a detailed statement, and distribute it for comment to other federal agencies having jurisdiction by law or special expertise, and to state and local agencies. If those agencies submit comments, the sponsor agency is to give them due regard and where necessary or desirable, secure modifications of the project that will minimize or eliminate environmental damage. Key features of the HUD environmental assessment process can be described by the following terms: decision points, normal environmental clearance, special environmental clearance, and preparation of the draft and final environmental statement.

Decision Points

The requirements of Section 102(2)(c) are aimed at obtaining full environmental consideration of a "*proposal* for legislation and other major Federal actions," or in other words, *prior to decision* by the agency. In Section 10, the Council on Environmental Quality (CEQ) guidelines discuss the stage(s) of program activity to which the guideline requirements will apply, remarking that each agency must determine those stages for its programs.

Generally in HUD programs the "decision point" is the approval of an application; however, because of variations between procedures among HUD programs, Appendix A of the draft Circular 1390.1 prescribes a "decision-point" for each program by which time "all HUD environmental clearances must be completed."

"Normal Environmental Clearance"

All applications, and all amendatories and changes to ongoing projects, must be given an environmental review in sufficient depth to permit the following determinations:

1. Project is below threshold and will involve no significant impacts—adverse or beneficial—on human environment and will not be controversial or precedent setting.
2. Project is below threshold, *but* review shows reasons to expect it to have "significant impacts" and/or prove to be environmentally controversial or precedent setting.
3. Project is above threshold.

The only exceptions to the foregoing guidelines are as follows:

1. Single-family home mortgage cases.
2. Applications, and so on, which are to be rejected or disapproved without further processing.

Such "normal review" will provide factual basis for determining whether: (*a*) the project is over threshold; (*b*) the project is under threshold, *but* appears to involve "significant impacts," and/or seems likely to prove environmentally controversial and/or precedent setting.

Thus if a case is found to be under threshold and does not appear to involve the environmental complications mentioned, a determination will be made on the review report that Section 102(2)(c) does not apply; thus no further environmental processing is required, and the case may go back into normal processing.

"Special Environmental Clearance"

Special environmental clearance involves a more complete, in-depth analysis of the environmental aspects of a case for which the "normal environmental review" shows a *possibility* that full processing under Section 102 is needed, or that the case is over threshold; in either event, the full analysis is required.

The analysis identifies and evaluates all environmental factors involved; thus we have a factual basis for determining whether full processing under Section 102(2)(c) is needed.

Regardless of whether the project under analysis is over or under threshold, if the analysis shows no reason to expect "significant impacts," environmental controversy, or the setting of precedents, a determination will be made in writing that Section 102(2)(c) does not apply and that the full processing under that Section is not required; the case then returns to normal processing.

Conversely, if analysis indicates reason to expect the project to have significant impact, or to be likely to cause environmental controversy or precedent setting, a draft environmental statement must be prepared.

In making this analysis, staff will use all available material from the applicant and from other sources, including information submitted by local groups opposed to the project. In these evaluations, all appropriate skills available in HUD Area and Insuring offices can be called on.

The Urban Environment—HUD's Role

In accordance with the law, HUD will endeavor to comply with the provisions of NEPA to "the fullest extent possible," rather than the extent feasible, or the extent practicable or convenient. Where the physical environment is affected, the problem is likely to be definable and probably quantifiable, and a solution can usually be developed. However, when the "human" or the "urban" environment is affected, and the "receptor" of the adverse effect is—immediately or secondarily—the city residents, impacts are hard to define and harder to measure, and the damages caused (e.g., to people's mental health) are well-nigh impossible to "cost." Much of the seeming difficulty probably will be reduced as experience is gained by HUD and by other departments, as well as by industry and science. Meanwhile, HUD—like other federal agencies—must do its best to explore and solve the environmental problems inherent in, or caused by, urban matters.

Aesthetic Assessment Methodology

MARILYN D. BAGLEY

> Aesthetics is concerned with the characteristics of objects and of the human being perceiving them that make the objects perceived pleasing to the senses.

The Problem

The practical concern for aesthetic considerations by project planners is twofold: to ensure the ultimate success or acceptance of the project by the surrounding community, and to meet the requirements stated in regulatory guidelines for publicly funded projects. The first concern usually initiates an honest attempt by the project planner to consider the visual impact of the project. This approach typically leads to obtaining expert advice from experienced designers in the area. Much to the bewilderment of the planning team, however, the "cosmetic attempt" does not always succeed in making the project acceptable to the community.

The second concern, that for meeting regulatory requirements, is the one currently receiving considerable attention. The term "aesthetics" appears in many of the federal and local guidelines for project planning, including the Council of Environmental Quality (CEQ) guidelines for responding to the National Environmental Policy Act (NEPA). This inclusion is a response to increasing concern in many areas of our country for the problem of making rational decisions when the tradeoffs include factors that are difficult to quantify.

Ms. Bagley is employed by Stanford Research Institute as an Operations Analyst.

Regardless of the motive for aesthetic concern, the real problem lies in the need for further development of the state of the art of measurement and understanding of aesthetic factors. In recent years, some attempts have been made to both quantify and qualify aesthetics, thus permitting the aesthetic aspects of a proposed project to be evaluated concurrently with the engineering, social, and economic aspects.[1-3] Such attempts have been made particularly in the areas of architecture, water resource, transportation, and public utility planning. A major difficulty, however, has been the lack of comprehensive systematic methods for evaluating aesthetic factors during the project planning stage. Part of this difficulty has been due to uncertainty regarding public perception of and concern for aesthetic values. If project planners had a way to determine the areas that would have aesthetic impact, and if they could make design adjustments prior to project presentation, many objections could be avoided. Moreover, where an impact could not be avoided, attempts could be made to justify it on the basis of the tradeoffs considered.

A concept of a methodology to make an objective assessment of the aesthetic effects of a proposed project has been developed[4] and applied in a case study[5] funded by Stanford Research Institute. For this case study, one of the proposed extensions for the Bay Area Rapid Transit (BART) system, the San Francisco Airport Access Project, was chosen as the prototype project. The objective of the case study was to show how a methodology could be applied to give structure to an aesthetic assessment. The prototype project was explored in sufficient depth to demonstrate the usefulness of the assessment procedure for identifying project impacts of aesthetic relevance and to outline how the resulting information could be applied in the evaluation of alternatives, the process of obtaining feedback from the public, and the preparation of Environmental Impact Statements.

Assessing Aesthetic Impact

Approach

Methodology denotes a step-by-step procedure used in a study, to provide a comprehensive framework for analysis. Three basic steps are used in assessing environmental impact: *identification* of the impact in terms of the physical and biological effects that cause it, *quantification* of the impact in terms of the dimensions of the physical and biological effects and of the number of individuals and the social entities that are affected, and *evaluation* of the impacts in terms of the attitudes and interests of those affected. These steps can be described for an assessment of aesthetic impacts as follows:

- *Identification.* The objective of the procedure developed for assessing aesthetic impacts is to provide information concerning the proposed project and its impacts, both direct and indirect, on aspects of the environment that might have aesthetic value to people in the area. The incidence of impacts depends on the geographic area considered, which is arbitrarily defined by the project staff prior to analysis.[6]

Direct impacts are those transmitted through the senses, resulting in a physiological or psychological response to a stimulus. Indirect impacts result from interpersonal relationships and social pressures.

- *Quantification.* The identifiable impacts are quantified by determining insofar as possible the magnitude of their effects—for example, the proportional number of people who would be directly or indirectly affected by an increase in noise, a visual barrier, a noxious odor, or a commercial or industrial development, or by the elimination of the open space, a cluster of trees, or a natural stream in an area.
- *Evaluation.* The information assembled from identification and quantification allows project staff to consider aesthetic impacts along with economic, social, and ecological impacts when evaluating tradeoffs among alternative designs for the project. Public feedback is a key element in the evaluation process.

Identification. Aesthetics is concerned with the characteristics both of objects and of the human being perceiving them that make objects pleasing to the senses. Thus in identifying the relevant characteristics of the project, the environment, and man, the aesthetic assessment methodology attempts to show how these elements relate to one another to enable man's aesthetic needs to be met. The methodology is based on the assumption that man's aesthetic needs can be expressed in three categories: *physiological*, as perceived by the senses; *social*, often determined by peer groups or other individuals; and *psychological*, related to emotional needs (see Figures 1 and 2). Many of these needs are amenable to quantification—for example, the amount of decibels (the noise level) consistent with human comfort and the atmospheric conditions necessary for good visibility. Others can be defined—for example, the need for tranquil space, together with the uniqueness of a particular area serving this purpose. By describing these needs, we can often show the scale of environmental requirements (ambient levels) to satisfy them.

By a similar breakdown of the physical environment pertaining to a specific project site and its immediately surrounding area, the methodology demonstrates how elements of the environment relate to man's needs. It also points out possible sources for background information and data concerning these elements. The changes expected to result from the proposed project are depicted by describing the environment without the project and by predicting what it would be like with the project, with special emphasis on the elements having aesthetic attributes.

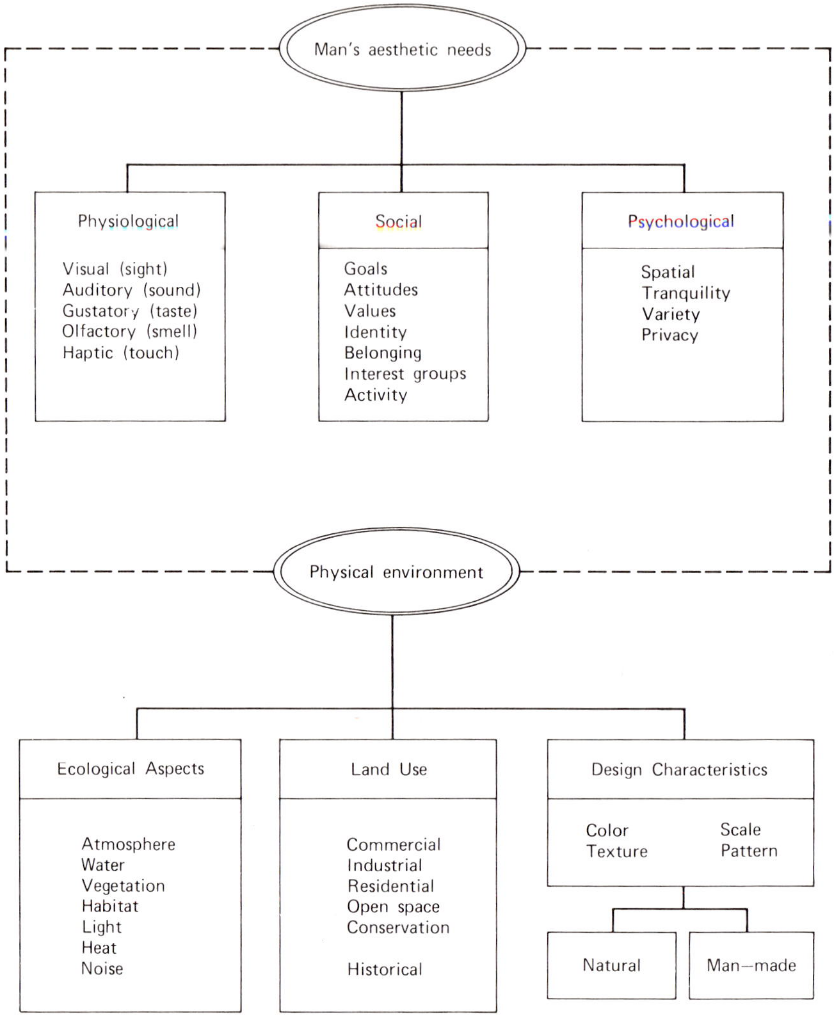

Figure 1. Identification of aesthetic variables.

A series of two-dimensional matrices has been developed to show the potential interaction of aesthetic needs with project effects (Tables 1, 2 and 3). The matrix procedure is a way of developing checklists identifying the interactions; it does not show the nature of an interaction nor its extent, since these features would be specific to the project under consideration. The first matrix is designed to identify areas in which man's physiological needs are affected by the environment. The variables used to identify man's physiological needs are based on the human senses and are not easily changed.[7-9] The second matrix is designed to show areas in which man's social and psychological needs are affected by his surroundings. The variables used to identify social and psychological needs are more flexible and should respond to the values of the people in the community being studied. Some experimental sampling may be required, followed by adjustment of the variables used. The social and psychological variables used in the case study were selected on the basis of a review of current articles describing man's hierarchy of needs[10, 11] and dis-

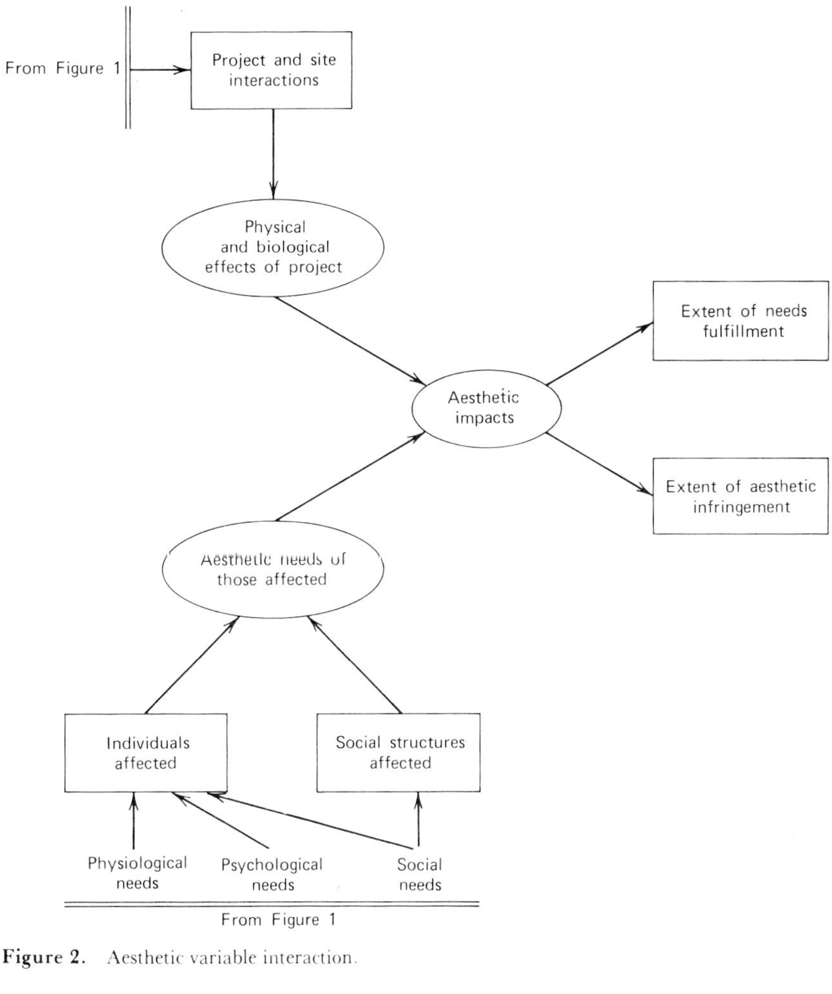

Figure 2. Aesthetic variable interaction.

Table 1. Matrix of Man's Physiological Needs in Relation to Ecological Elements

Ecological Elements	Physiological Needs[a]			
	Visual (sight)	Auditory (hearing)	Olfactory (smell)	Haptic (touch)
Atmosphere				
Air quality (gases, particulates)				
Climate				
Temperature				
Water				
Quality				
Temperature				
Type (ocean, stream, pond)				
Earth				
Composition				
Landform				
Vegetation				
Species (crop, trees, shrubs)				
Barriers				
Corridors				
Habitat				
Wildlife				
Birds				
Fish and shellfish				
Insects				
Benthic organisms				
Microfauna, fauna				
Light (artificial and natural)				
Sound source (noise)				
Man-made				
Natural				

[a] Above diagonal line represents "before" project, under diagonal line represents "after" project. No mark denotes average or ambient conditions (—) equals negative, (+) equals positive impact.

cussions with staff sociologists. A third matrix identifies areas of potential impact on the physical environment by the project itself.

Quantification. To use the matrix checklist, the project evaluator can go through each matrix and analyze the relationship between the two variables, determining on the basis of familiarity with the project and the environment, where an impact would occur. Only interactions showing effects of some defined magnitude are depicted. A diagonal line in each square separates the impact "without" the project from the impact "with the project." Positive and negative effects are indicated by using a plus or minus sign in each square. These could be rated further by assigning numerical weights (from 1 to 10), or high, medium, and low rankings if some standard mean were established for weighting factors. Following the matrix, each element showing an impact is described in text form.

The proposed BART extension project was divided into physical segments for analysis, each segment having a complete set of matrices for aesthetic impact identification. Most projects requiring environmental impact analysis are more limited than the BART project in the amount of geographic space they encompass, therefore would not require division into segments. Transportation projects (including highway planning), however, affect the entire corridor of the route alignment, traversing many diverse sections of the environment. Analysis by segment is a comprehensive way to depict impacts along transportation corridors. This method assumes that the impact of the whole is no greater than the sum of the parts. This may not always be the case, but is a useful first approximation.

After impacts for each BART segment had been identified and some means of depicting the magnitude of the impacts had been assigned, the entire project route was analyzed by summarizing the significant impacts of the various segments. This method was designed to facilitate comparison of alternative route alignments; more important, it identified within a proposed route the problem areas that could be remedied by a design adjustment (e.g., for noise, a buffer wall could be added; for invasion of privacy, a different track elevation could be considered).

Impacts were identified for both short-term and long-term effects. When completed with the necessary data for quantifying, the assessment will provide a basis for comprehensive review of the proposed project and will also help the BART planners to identify various design alternatives that might lessen impacts.

Evaluation. After various disciplines (social, biological, legal, economic, air, and water quality) have assessed impacts pertaining to their respective specialities, impacts are considered as they relate to each other. This enables an analysis of secondary-order impacts that might occur and often points out a chain of effects that might not have been anticipated in individual analysis.

Table 2. Matrix of Man's Social and Psychological Needs in Relation to Project Site

Project site	Social and Psychological Needs[a]											Design Characteristics[a]				
	Spatial	Tranquility	Variety	Privacy	Educational	Economic	Political	Interest Group	Identity	Activity	Occupational	Scale	Color	Texture	Pattern	Other
Land use																
Residential																
Industrial																
Commercial																
Open space																
Conservation																
Recreation																
Man-made features																
Transportation																
Utilities																
Barriers																
Energy source																

Natural features																
Water																
Vegetation																
Habitat																
Landform																
Special features																
Historical																
Unique																
Other																

[a] Marks above the diagonal line represent "before" the project; marks under the diagonal represent 'after' the project. No mark denotes average or ambient conditions; (−) represents negative impact; and (+) represents positive impact.

Significant impacts are then outlined in an interdisciplinary way for each design alternative, including the alternative of no project, and evaluation follows.

A variety of evaluation methods are used in the planning process; it is beyond the scope of this chapter to go into these in depth. One method for which the aesthetic assessment methodology is particularly suited is that of "general systems theory." This procedure evaluates alternative designs on the basis of how well they satisfy carefully defined community and project objectives. Evaluation criteria serve as a tool for judging the various factors and for assigning numerical weights for comparison. The alternative best satisfying the problem statement within the stated objectives is the design selected for consideration.

In this method of evaluation and in many other methods, the use of some type of feedback is considered a necessary element in planning. Feedback usually comes from the community involved. It can be obtained directly (by public meetings, surveys, votes, or polls of interest groups), or it can be obtained indirectly by analysis of social indicators (voting trends, letters to the editor, public records, lawsuits, etc.).

Feedback is most important in environmental impact evaluation, for determining desirable tradeoff priorities. Reliable feedback often requires prior public education. Tools for assessment and impact identification can be useful for describing impacts in meaningful ways (nontechnical terms) to the general public. Graphic presentations, photographs, and benefit/cost figures are especially useful for facilitating information transfer and for stimulating response. These feedback tools have been incorporated in the proposed aesthetic methodology.

In the final analysis, consideration of the costs and impacts of a proposed project, together with its value to individuals and the community, requires the exercise of judgment. Since the provision of objective information is vital to well-informed and wise decisions, the research community must begin providing methods for improving the state of the art of measurement and identification of all elements of environmental concern. The methodology for aesthetic assessment is one such attempt; it now needs to be applied to various planning procedures and analyzed for the policy implications it may generate.

Notes and References

1. L. Leopold, "Landscape Esthetics, How to Quantify the Scenics of a River Valley," *Natural History*, **78** (October 1969), 36–45.
2. D. Appleyard and R. Okamoto, "Environmental Criteria for Ideal Transportation Systems," Institute of Urban and Regional Development, University of California, Berkeley, 1968.

3. R. B. Litton, "An Aesthetic Overview of the Role of Water in the Landscape." Prepared for the National Water Commission of the Department of Landscape Architecture, University of California, Berkeley, July 1971.

4. M. D. Bagley, "Aesthetic Assessment Methodology for Environmental Impact Analysis," Technical Note TN-OED-004, Stanford Research Institute, Menlo Park, Calif., March 1972.

5. M. D. Bagley, "Aesthetic Assessment Methodology—A Case Study," Technical Note TN-OED-011. Stanford Research Institute, Menlo Park, Calif., November 1972.

6. For the purposes of the case study, the geographic area of primary concern was within a quarter mile of the proposed project site.

7. J. M. Fitch, "Experimental Bases for Aesthetic Decision," in *Environmental Psychology: Man and His Physical Setting*, H. M. Prochavsky et al., ed. Holt, Rinehart, Winston, New York, 1970, pp. 76–84.

8. *Handbook of Human Engineering Data*, Tufts College, Institute for Applied Experimental Psychology, 1951, Part I, Chap. III, Sec. 1.

9. H. Lee, *Perception and Aesthetic Value*, Prentice-Hall, Englewood Cliffs, N.J., 1938, p. 100.

10. D. W. Prall, *Aesthetic Judgment*, Thomas Y. Crowell, New York, 1929, pp. 56, 67, 310.

11. B. Berelson and G. Steiner, *Human Behavior—An Inventory of Scientific Findings*, Harcourt, Brace, Jovanovich, New York, 1964.

An Ad Hoc Environmental Impact Statement

MARY A. JAROCH

The National Environmental Policy Act

The National Environmental Policy Act requires that *all* federal agencies implement its provisions to the *fullest* extent possible.

I have italicized two words to emphasize the all-inclusiveness of the law, which has successfully weathered attack by way of the courts and the executive branch by agencies wishing to avoid this duty. Bills have (without success) been introduced in Congress for agencies such as the Atomic Energy Commission, attempting to obtain blanket exemptions for all agency pursuits.

Furthermore, the courts have held that units under federal contract to various agencies are not exempt: that is, the agency awarding the monies must file an environmental impact statement (EIS) for any action undertaken by its subcontractors. This prevents bureaucratic buck passing and relieves the subcontractor of the ethical onus of inadvertent environmental decision making.

Such an equitable arrangement, one might think, would lead to understanding, enlightenment, and all the good and just items mentioned in Section 101 of NEPA:

> Section 101. (*a*) The Congress, recognizing the profound impact of man's activity on the interrelations of all components of the natural environment, particularly the profound influences of population growth, high-density urbanization, industrial expansion,

Ms. Jaroch is an Environmental Specialist with the College of Engineering and Applied Science, University of Wisconsin, Milwaukee.

resource exploitation, and new and expanding technological advances and recognizing further the critical importance of restoring and maintaining environmental quality to the overall welfare and development of man, declares that it is the continuing policy of the Federal Government, in cooperation with State and local governments, and other concerned public and private organizations, to use all practicable means and measures, including financial and technical assistance, in a manner calculated to foster and promote the general welfare, to create and maintain conditions under which man and nature can exist in productive harmony, and fulfill the social, economic, and other requirements of present and future generations of Americans.

(*b*) In order to carry out the policy set forth in this Act, it is the continuing responsibility of the Federal Government to use all practicable means, consistent with other essential considerations of national policy to improve and coordinate Federal plans, functions, programs, and resources to the end that the Nation may:

(1) fulfill the responsibilities of each generation as trustee of the environment for succeeding generations;

(2) assure for all Americans safe, healthful, productive, and esthetically and culturally pleasing surroundings;

(3) attain the widest range of beneficial uses of the environment without degradation, risk to health or safety, or other undesirable and unintended consequences;

(4) preserve important historic, cultural, and natural aspects of our national heritage, and maintain, wherever possible, an environment which supports diversity and variety of individual choice;

(5) achieve a balance between populaltion and resource use which will permit high standards of living and a wide sharing of life's amenities; and

(6) enhance the quality of renewable resources and approach the maximum attainable recycling of depletable resources.

(*c*) The Congress recognizes that each person should enjoy a healthful environment and that each person has a responsibility to contribute to the preservation and enhancement of the environment.

Instead, however, Murphy's law has swung into play, and everything that could possibly go wrong, has. NEPA Environmental Impact Statements offered by agencies are all too regularly produced by the very engineers designing the proposed project, without the assistance of objective ecologists and other professionals, the funding for original research uncontaminated by special interest funding,[1] or any evidence of community self-determination.

At best, the product is shabby boiler-plate boredom, at worst, a self-serving, biased, and sometimes falsified "document" justifying environmental exploitation for economic gain.

A General Accounting Office Report analyzing the performance of seven federal agencies in preparing EISs found that:

. . . the requirements of Section 102 are not being carried out uniformly and systematically.

Most of the seven agencies did not:

- complete environmental impact statements in time to accompany proposals through all agency levels of review,
- complete the statements in time for them to be used in the early stages of decision making, and
- review effectively the results of plans to ensure that the environment is protected as anticipated.[2]

An Ad-Hoc Effort Begins

Thus when we were asked to produce an ad hoc EIS on the proposed Park-West Freeway through Milwaukee in three months, with almost no funding, our trepidation was understandable. If so many "experts" with so much money and numerous established organizations could not produce an EIS to conform fully to federal law, how could we, a specialist and a team of students, possibly hope to emerge victorious? On the other hand, we had all read reams of inadequate "official" EISs and we could hardly do worse; at least, we knew what *not* to do.

In support of our efforts, there was an urgent public need to be filled. Judge John Reynolds of the U.S. District Court, Eastern District of Wisconsin, had granted a preliminary injunction against Park-West at the suit of a citizen's neighborhood coalition pending the acceptance of a viable EIS:

> "It is far more consistent with the purpose of the Act (NEPA '69) to delay operation at a stage where real environmental protection may come about than at a stage where corrective action may be so costly as to be impossible." . . . Failure to grant a preliminary injunction at this time would allow further investment of resources into Park-Freeway West, making its abandonment or alteration in light of environmental concerns increasingly costly and increasingly difficult. Indeed, it might even cost plaintiffs their case, for if construction continues it might well reach the stage where NEPA would be inapplicable.[3]

The highway agencies concerned let it be known that it would take at least another year to prepare the official EIS. Indeed, that was one of their arguments in court, and in February 1973, no EIS was in sight.

Public need also compelled the U.S. Department of Housing and Urban Development (HUD) to state that *no* further freeways could proceed until the more than 2000 units of housing already destroyed were replaced, especially for low-income families. This requirement virtually rules out further freeway land acquisition and/or demolition, since Milwaukee's housing programs are at a political and financial impasse.

Third, state highway Department officials had already reallocated highway building funds for Park-West and the court-aborted Lake Freeway[4] to other areas, primarily the up-grading of existing highways throughout the state.

Therefore, an informational gap about the impact of the proposed Park-West had widened into a virtual abyss of legal, environmental, and economic confusion. The citizens of the community surrounding the stripped swath of right-of-way were actively opposing freeway construction with little in the way of substantive fact, since the only studies available—those by Howard, Needles, Tammen, and Bergendoff, the consultant design and engineering firm—did not include the factors most important to the citizens concerned.

Instead of compiling a monolithic, obtuse, and impassive report, we decided that the first six goals of Section 101 would guide our research (and the use of it) to form an educational, economic, social, and environmental background that any interested taxpayer could readily utilize to make a decision on the issue. We were less interested in producing an "academic" document than a useful one that internalized "social" costs to the community and facilitated self-determined neighborhood action. Thus although accuracy was in no way sacrificed, jargon, overly technical description, and ivory-tower abstractions were avoided in favor of those values the community had not received and urgently requested: housing, community renewal, economics, and environmental quality of life. After thoroughly delineating first the present neighborhood, and second, the proposed freeway and its impact, our aim was to offer a generalized alternative to building the freeway as a possible option to be decided on by the affected community.

Methods: Writing an EIS in Spare Time

The very first resources required in EIS writing are the all-important human ones. Several former students who came from disparate disciplines and were highly motivated: an architecture student, a biology student now in graduate computer work, and an engineering student employed for the summer by the state Department of Transportation comprised the basis research team and did most of the writing. They divided the work among themselves, kept notebooks, devised questionnaires, and went from door to door interviewing residents and small businessmen of the area. Weekly sessions were held in which the progress of the work and any developing gaps or new needs to be filled were discussed.

Graphic Content

Meantime, as the text took shape, graphics to clarify all the statistics became necessary. For an informational document, maps were an absolute necessity. Some, showing subpopulations (children of school age, elderly people, ethnic groups) of the area, could be quickly drawn over standard forms.

Others, the essential, doublefold maps of the entire freeway right-of-way, were done by an environmental architect (Robert Garcia) who volunteered original drawings of the following:

1. The present cleared right-of-way and the neighborhood composition: every residence, church, school, street, and alley was shown.

2. The proposed Park-West Freeway as it would look if built to present plan, with off-and-on ramps, depressed grading, tunnels, overpasses, and so on.

3. The architect's alternative plan, which presented an ingenious modular concept including privately built and owned housing for small or large families and elderly people in single-dwelling, apartment, or townhouse form. Mixed in were block-sized modules of recreational urban park, new stores, and community service buildings such as a well-baby clinic, Social Security office, and day care center.

Instead of a ten-lane concrete freeway, he indicated a two-lane tree-sheltered bus transit line for express or local service, feeding the neighborhood and the downtown area.

Diversity, the great law of ecological balance, was the assumption on which the alternative was based, and community self-determination was the freedom it offered. If implemented as is, the plan would replace 700 units of housing for all incomes and ages.

Needless to say, the EIS would have been far less effective without these maps, for on each one a citizen could pick out his own house, church, or office, and relate it to the right-of-way.

Computer Use in EIS Formation

A unique contribution was offered by James Otto, another architecture student, who devised a 300-slide computer pictogram of the proposed freeway. Each slide represents a view of the road as seen from a car. This invaluable educational tool allowed the slide viewer to "drive" east or west on the freeway and "experience" it before the first concrete was poured.

One transportation engineer described an existing Milwaukee freeway, which has become synonymous with poor design and safety as follows: "It was planned as a series of 'events': each interchange was an event and designed without the previous or the following one in mind; no thought was wasted on the miles of road between the events."[5]

Raison d'Être and Like Matters

Whenever the team found itself mired in statistics, maps, questionnaires, footnotes, and graphs, a review of the primary *raison d'être* of the EIS came to the rescue: the *Ad Hoc Environmental Impact Anaylsis of Park-West* was conceived, implemented, and produced to *inform the public.* Whenever obstacles presented themselves, they were easily dissolved by invoking this reminder. And, again, we could ask ourselves: What should *we* do that existing EIS's have not?

Here are some of the "other" things we did:

Community Cooperation

Several excellent community organizations had coalesced around the Park-West issue before we began our work. Instead of evading and avoiding them, we participated in many of their meetings, being as open as possible about our work. Heresy! Some may shout, How can we support claims of objectivity when we encourage close communication and cooperation with little-old-ladies-in-tennis-shoes? Easily. We recognized that for our purposes both political entities and citizens had to be consulted. A one-sided view could not suffice. It was not possible for us to accept the belief that knowledge and wisdom could exclusively reside, for instance, in the Expressway Commission, or the mayor's office. Wisdom, if we are to believe our founding fathers, is to be found also, if not primarily, in the expressed will of the people.

Second, this "expressed will' is nowhere better informed or more genuine than in self-propelled citizen interest groups. They have no billion-dollar budgets to defend, no elections to win, no political axes to grind with any issue that comes to hand. They *do* stand to *lose*: their houses, neighborhoods, schools, churches, and their constitutional right to the pursuit of happiness. It is this right which has been consistently, even ruthlessly, ignored in direct contravention of the stated intent of NEPA 1969. I think, in fact, that an EIS that does *not* work with citizen organizations is a defective statement de facto.

As the GAO report concludes:

> . . . [agency] efforts to achieve public participation seem less than intended by the act and the Council's guidelines.
>
> Agencies need to take an active role in public meetings, use environmental impact statements as an integral part of the meetings, and perhaps most important, experiment with innovative concepts for improving communications with the public on environmental impacts.[6]

and again:

> The Council [on Environmental Quality] reported that individuals and groups often could contribute data and ideas beyond the expertise of the agency involved, that citizens were seeking and making significant changes in agency policy, and that citizens often could learn of an impending action or the environmental issues raised only by way of an environmental statement.[7]

Unfortunately, the agencies most in need of greater communication with the public are often extremely defensive, even to placing themselves unnecessarily in an adversary position to those paying for their work. Additional bad feeling arises when the *validity* of public opinion is repeatedly called into question, with epithets such as: eco-freaks, L.O.L.'s, bird-watchers (sic), and radical nuts, *ad nauseaum*.

If, on the other hand, care is taken to initiate *nonmanipulative* interaction with the neighborhood, surprising help is often forthcoming. Honestly solicited information is readily given.

Of course, we all known how to *dis*honestly solicit opinion:

- Wife-beating: "Are you for the freeway, or do you want to see the city *lose* x-million dollars?"
- Abstractions: "Do you think the freeway has a bad/good effect on the economy?"
- The "one-child" argument, so dog-eared with use: "If we save the life of *one child* by widening this road (to make it safe is the implication), all and any cost will be worth it."

The public is no longer, if they were ever, taken in by such transparent machinations, and bitter resentment of such patronization may escalate into court or street action.

If, however, specific information needs are made plain, and the team did this, the response will be ready, if not preordained.

Questions like: How many people does your business employ compared to last year? Will you be hiring more next year? Is your net profit up or down since last year? can reflect true economic gain and loss.

Questions like: Do you drive or own a car? Do you work downtown? Would you use this new freeway? and, Do you take the bus? Where to? If not, why not? bring mass transit needs into sharp focus.

Finally, quality of life becomes the issue. Do you have children? How many? Do they play (*a*) in your yard (*b*) in the street, (*c*) at a park? Is this a safe neighborhood? All the time? When is it not safe? Are you going to stay here? If not, why not?

Patience in this area is indeed its own reward. In addition, these separate factors add up in a new light: people who thought environmental protection meant bald eagles *experience it* firsthand as meaning themselves and their children. No more pernicious myth exists, and kicking these decisions into a boondocks court or to "the experts" only perpetuates it and increases the conviction of helplessness and lack of self-worth on the part of many citizens.

Incidentally, this issue brought together people of many colors, convictions, and levels of formal education. Individuals from diverse ethnic backgrounds found that they could easily agree over the issue, and proceed to act. The Ad Hoc EIS provided "expert" fuel for them: it *served* them in a new way. If for no other reason, its existence is justified.

Epilogue: Political Animals

Although we kept a "low profile" politically during the earlier stages of preparation, rumors of the report reached many corners of the city.[8] A copy of the alternative plan map reached the mayor's office through carelessness, and he abruptly announced against Park-West and for housing, two days before the Ad Hoc EIS was officially published.

News people were given some of the first copies, and extensive play in the press and television ensued. The sponsor of the work, Congressman Henry S. Reuss, who had funded and encouraged us, decided to hold congressional grass-roots hearings on Park-West and Stadium North Freeway November 14 and 15, 1973. These in turn spawned more controversy and turned up more facts and opinions.

For those who decry the political "circus," I would like to present the alternative: privately arrived-at decision making by those in power. For those who know hearings are, in the legal sense, impotent, I commend the courts as a way of establishing precedent firmly.

If, however, we aim to educate, to facilitate, to serve the public interest, we must depend, indeed, *insist* on the involvement of mass media at our conferences and doings.

The returns on Park-West are not in yet. In spite of the large legal obstacles to its construction and the financial problems, there is an ongoing controversy: shall we have a referendum (in contravention of NEPA 1969)? If so, shall the county hold it? Shall the city hold it? (The Ad Hoc EIS showed that suburban residents, not city dwellers, would benefit from the Park-West.) What are we going to do about housing replacement? And so it goes. *But the EIS was done.*

Notes and References

1. The seismological research prior to the leasing of the Santa Barbara oil drilling platform which "blew" was done for U.S. Geological Survey by the two oil companies building and retrieving the oil.
2. Improvements Needed in Federal Efforts to Implement the National Environmental Policy Act of 1969 (B-170186), Report to the Subcommittee on Fisheries and Wildlife, Conservation Committee on Merchant Marine and Fisheries, House of Representatives, May 18, 1972, Washington, D.C., pp. 1, 2.

3. Decision and Order, Civil Action # 72-C-300, *Northside Tenants' Rights Coalition et al.* vs *John A. Volpe, Secretary, U.S. Dept. of Transportation*, July 30, 1972, p. 10.
4. An unsatisfactory EIS was filed on the Lake Freeway and rejected by the courts and EPA as inadequate. Also, a citizens' group successfully brought suit against the expressway as a violation of a deed restriction on the use of the park land involved for anything other than park and recreational uses. *Roger Cobb, John Ong, Jr., Mrs. John Flynn, T. K. Phillipson, Dan Cupertino, Russell Knetzger* vs *Milw. County Expressway and Transp. Comm., Wis. D.O.T. and City of Milwaukee*; Wisc. State Sup. Ct., Case # 461.
5. Private conversation, Dr. Edward Beimborn, University of Wisconsin, Milwaukee, 1972.
6. *Op cit.*, Improvements needed in Federal Efforts to Implement the National Environmental Policy Act of 1969 (B-170186), p. 33.
7. *Loc. cit.*
8. Low profile was desirable to avoid locking in the typical agency defense mechanisms, and in short, cutting off one source of information. Also, people might color their responses to us unnecessarily.

Part III

Environmental Indices and Environmental Management

Terry Davies

Mr. Davies, formerly Senior Staff Member on the President's Council on Environmental Quality, now is with Resources for the Future.

It is quite appropriate that a conference devoted to environmental management should include a discussion of environmental indices. Good measurement and good management are closely related. It is no coincidence that the area of public policy for which we have the best measurements and indicators, the national economy, is also the area in which we have been most successful in using the tools of government to achieve public goals.

The satisfactory carrying out of many basic management functions depends on the development and use of indices. The establishment of priorities, the provision of information to decision makers and the general public, and the evaluation of ongoing programs all would be vastly improved if greater use were made of indices. One wonders, for example, how effectively we can evaluate the success of our multibillion dollar water pollution programs when we do not have good overall measures of whether water quality is improving or getting worse.

The Council on Environmental Quality has been working for almost two years on the development of environmental indices. I believe we have made progress, but the task is complex and often frustrating. Often it seems that when we know what we should be measuring we do not have good data, and when we have good data we do not know the right way to put them together.

If an adequate system of environmental indices is to be assembled and periodically published, it will require the efforts of many agencies and individuals. The contributors of the Chapters in Part III are representive of those at the forefront of attempts to gain acceptance and utilization of environmental indices. To the degree that they are successful, environmental management in this country will be markedly improved.

Environmental Indicators for San Diego

D. G. MALCOLM

DONALD MACDONALD

Abstract

Under the auspices of the Integrated Regional Environmental Management program in the San Diego region (Project IREM), work has been undertaken to develop a series of Environmental Quality Indicators for use in communicating the state of the environment to citizens and decision makers in the region. The work has been undertaken in two phases. The first phase, completed in June 1972, defined the state of the art nationally and locally in measurement of environmental quality and recommended a plan for development of an Environmental Quality Information System. In all, some 12 major indicators of environmental quality were defined, and each was deemed useful and capable of being computed. These were grouped in four major categories—Air and Water Related, Land Related, Multimedia Related, and Social Aesthetic Related.

Implementation of Phase II began in October 1972 with an effort directed toward development of the following indices: Air, Water, Noise, Energy, Solid Waste, and Land Use. These areas have been deemed to be of primary concern and importance in the region.

Over the past 20 months, Project IREM has directed significant effort to the objective of bettering environmental decision making through improved, more timely and integrated information concerning the regional environment. In

Mr. Malcolm is Dean of the School of Business and Economics, California University at Los Angeles. He and Mr. Mcdonald, of General Analysis Corporation, worked on Phase I when they were with Research Analysis Corporation.

furthering this objective, IREM has completed and published a nine-volume "Natural Resources Inventory" and a five-volume series of Regional Issues, one of which defined the need and role of a system of Environmental Quality Indicators (EQI).

One of the most basic questions that immediately arose involved the need to have a frame of reference within which to set objectives and to measure the performance of the many programs designed to enhance or maintain the quality of life in the region. It appears that if one is to influence the factors affecting the environment, one must know which factors and how to measure when the environment has indeed changed. Ideally, the physical characteristics should be measured periodically. If these characteristics were monitored over time and correlated with the activities of environmental control programs, program effectiveness could be evaluated and problem areas pinpointed.

The problems of determining the right factors to be included in such an index presented major problems. The Council on Environmental Quality (CEQ) had listed 112 indices that measure some aspects of the quality of life,[2] and over the years some 56 monitoring programs have been created by various federal agencies. Over and above these are important indicators of land use expectations coming from the private sector, such as the developers and the financial community. The capability of forecasting trends in environmental factors is also desirable. The questions of what is important to measure, how valid the measurements are, whether the data are available, and just how specific data can be used in monitoring and forecasting trends are important to resolve.

To evaluate and resolve these questions, IREM initiated a multiphased program to develop a series of Environmental Quality Indicators for use in communicating the state of the physical environment to citizens and decision makers of the region via a state-of-the-environment report. Phase I was a feasibility study; the second phase of activity is currently underway in IREM to develop seven of the dozen indices of Environmental Quality recommended and to publish a state-of-the-environment report for the San Diego region.

Phase I Activity

Objective

To explore the state of the art in environmental measurement and monitoring, the relevant indices and indicators that are in use, or planned, to evaluate the San Diego region's special needs and capabilities in regard to environmental information, and to recommend an appropriate plan of action for enhancing the understanding of and communications concerning San Diego's physical environment.

Approach to Study

The study team conducted direct interviews of prospective users of the index. To determine what was practical, a thorough search was made of the literature and direct contacts were made with other individuals and groups that are doing work related to the development of an Environmental Quality Index (EQI). The study team experienced the same problems others have in attempting to select meaningful factors for use in assessing the state of the environment. We found Quality of Life (QOL) and Quality of Environment (QOE) to overlap at every turn. After much discussion, we have decided that the state of the *physical* environment—both natural and man-made—was the subset of QOL on which we would concentrate our efforts. We started with the following basic principles in selecting the indices:

1. An index must be of interest and value to the residents of the region and this includes government officials, special interest groups, and individual citizens.
2. It must be technically feasible to monitor the environmental parameters to be included in the index.
3. It must be economically feasible to accumulate and process the data needed to support an index of quality, either now or in the foreseeable future.
4. The parameters monitored and the indices used should reinforce rather than undermine any serious attempts that are being made to introduce standard or uniform terms for expressing environmental quality at the national level.
5. Both in meaning and format an index must be simple, precise, and easily communicable to citizens.

Both the availability and cost of data collection and the state of development of environmental controls were considered. It was found necessary for IREM to work with data supplied in the main by other operating agencies, within and outside the Environmental Development Agency. This stems from time and cost considerations as well as the fact that other agencies *are* responsible for such measurement and information. Therefore, a practical constraint on what can be developed as an index of the quality of the environment is the nature and availability of data from other regional agencies. Corollary to this is the need to determine and describe the current and potential interfaces of IREM with the programs of such federal agencies as the U.S. Environmental Protection Agency (EPA) and various state agencies that exercise control over regional regulatory bodies.

Availability and cost of securing data were found to vary widely from one environmental category to another. We determined that if a particular area of environmental concern is to develop into a "mature control technology," several evolutionary steps must occur. When all these steps have been accomplished, we have a completely viable control technology, complete with procedures, standards, and information for considering the legal, economic, and technical factors involved in environmental control:

1. Identification of pollutant/degradent in measurable terms.
2. Achievement of a practical means of detection/measurement.
3. Defined relationship of pollutant as a degradent to health, economic, or aesthetic values.
4. Specification of minimum standards for presence of pollutant or degradent in the environment.
5. Specification of permissible amounts of pollutant or degradent to be emitted per unit of time.
6. Establishment of implementation goals (i.e., maximum pollutants levels and/or maximum emission levels by all emitters).
7. Development of enforcement means that may require new legislation or administrative procedures.
8. Development of a means of monitoring and communicating the status of the environment and the results of regulatory agencies.

Using the "process" as just defined, the state of the art of the development of the 12 categories of environmental quality recommended for monitoring in the San Diego Region was made. Point values were assigned for each step, and values were summed to provide a rank order, or the relative maturity of each environmental category. These rankings are given below.

		Rating, 0–100%
1.0	Air and Water Related Categories	
1.1	Air Quality	100
1.2	Water Quality	85
1.3	Noise	70
2.0	Land Related Categories	
2.1	Land Use or Land Use Shift	40
2.2	Recreational Resources	45
2.3	Wild Life and Natural Vegetation	30
2.4	Natural Resource Depletion	30
3.0	Multi-Media Related Categories	
3.1	Hazardous Substances	55
3.2	Solid Waste	60
4.0	Social/Aesthetic Related Categories	
4.1	Housing Adequacy	35
4.2	Visual and Aesthetic Characteristics	30
4.3	Pests	55

Definitions of Terms Used

Because definitions of terms used presented problems in establishing precise and nonemotional communication concerning the purpose and use of environmental information, it is useful to state a series of definitions of terms that were used.

Goal. An environmental goal is generally a broad statement of desires relative to environmental quality. Current goals in areas may be: "not to further degrade" or "to meet use standards" for pollution emitters.

Standard. A standard is a law or a regulation with the force of law, specifying the amounts of contaminants permissible, or setting specific conditions that are allowed to exist. There are two types of standard:

Environmental. The maximum amount of contaminant considered tolerable in the environment—the amount permitted in ambient air or water.

Emission or Source Limits. The maximum amount of contamination or degradation permitted by an individual pollutant discharger or other agent of environmental degradation.

The Physical Environment. The physical environment for which we seek a measure of quality is that set of climatic, earth- and life-related factors, of which man is a part, that act on communities and organisms.

Quality of Life. Includes the condition of the physical environment and also considers the social, economic, and cultural factors that influence the life of individuals and communities.

Indicator. A signal or a symptom of the current condition of some aspect of the physical environment. We are seeking reliable indicators that are easily obtained measures of the current status.

Index or Indices. Ratios or other numbers derived from a series of observations of a particular aspect of the environment; as such, they are useful as measures of *change* in the environment.

Monitoring Techniques. The measurement, or monitoring, of the environment may be accomplished by a variety of techniques. These may often be used supplementally and are often alternatives offering different cost and accuracy of results:

1. Physical/Chemical.
2. Biological.
3. Economic/Statistical.
4. Social/Aesthetic/Resource.

Proposed Indices

The result of studies of regional and national interest led to the conclusion that there were twelve indices in four major categories which should form the basis for a San Diego Regional Environmental Information System. Tables 1–4 show the proposed indices.

Table 1. Environmental Quality Index Categories and Indicators for Air and Water and Noise Related Categories

Quality Index Categories	Definition of Category	Proposed Quality Indicators	Method of Expressing Quality	Data To Be Collected	Data Source(s)	Frequency of Producing Index	Proposed for: Initial System	Proposed for: Ultimate System
Air	Ambient air quality based on samples taken at 9 air monitoring stations in the populated section of the county, or based on level of contaminants in the air at the official monitoring stations, as opposed to emission levels.	O_x—oxidants SO_2—sulfur dioxide NO_2—nitrogen dioxide CO—carbon monoxide PM—particulate matter	Oak Ridge Air Quality Index—comparison of levels of contaminants versus California or federal standards	Monthly report from the two APCD stations that monitor all pollutants	San Diego Air Pollution Control District (APCD)	Monthly, quarterly, and annually	xxx	
		or O_x only (compared with Standard) or number of days/month that *any* pollutant is above the standard	Oxidant Index (worst hour average compared to standard)	Daily report from 6 stations that monitor O_x	San Diego APCD	Daily	xxx	
		Episode days per month or per year	Number of days/month or per year any standard is exceeded	Monthly report of all pollutants/month	San Diego APCD	Monthly		xxx
Water	1. Fresh surface water ●Lakes and reservoirs ●Recycled Water	●Dissolved oxygen (DO) ●Total coliform ●Nitrate ●Phosphate ●Temperature ●Total dissolved solids	Acceptable for each intended use: e.g., accepted for ●Drinking or Public Water Supply ●Shellfish and swimming	Level of each indicator in each type of water	1. Private water companies, Irrigation District and municipal water departments, controlling lakes and reservoirs	Semiannually	xxx	

	2. Ground Water	●Toxic substances ●pH ●Turbidity or transparency	●Swimming ●Noncontact sport ●Water fowl ●Agriculture ●Navigation		2. State Department of Water Resources	Triannually	xxx	
	3. Lagoons and bays				3. Regional Water Quality Control Board	Annually	xxx	
	4. Off-shore coastal waters				4. State Department of Navigation	Undetermined		xxx
Noise The EPA Noise Abatement and Control Office is currently working on a Community Noise Reference Scale that should assist in establishing norms and monitoring techniques.	1. Community noise, excluding industrial situations where workers can act to reduce their exposure	1. Ambient day, evening, and nighttime noise levels per upcoming EPA Standards	Decibels on the A scale, with index based on combination of decibels over Standard, factored by percentage of time Standards are exceeded.	Day and night noise levels in sample areas, with average, peak, decibel levels and duration and frequency of occurrence of peak noise levels.	1. Community noise to be measured by cities and county	Annual		xxx
	2. Airport area noise	2. Same per community noise exposure level (CNEL) guidelines			2. County for State of California	Quarterly	xxx	
	3. Freeway noise				3. State Department of Highways is monitoring noise in proximity of freeways.	Quarterly	xxx	
	4. New housing sites	4. HUD noise guidelines for new housing sites				Annual		xxx

Table 2. Environmental Quality Index Categories and Indicators for Land-Related Categories

Quality Index Categories	Definition of Category	Proposed Quality Indicators	Method of Expressing Quality	Data To Be Collected	Data Source(s)	Frequency of Producing Index	Proposed for: Initial System	Proposed for: Ultimate System
Land use or Land use Shift Acres converted to higher intensity use or converted from "open space" status to "developed" status (e.g., covered-over land)	Intended to measure rate at which land is being converted from low-intensity use; the rate at which "open space" is being used up, intrusions on ecologically sensitive or valuable land, and to provide a measure of "urban sprawl."	●"Urban green" or open space in urbanized areas	Total acres and percentage urbanized area	Acres added or lost each year	Regional Planning Data Base	Annual	xxx	
		●Intrusions into ecologically sensitive areas (e.g., wet lands, flood plains, agricultural land)	Total size and percentage of acreage intruded on: ecosensitive wet land, flood plain, or other	Acres intruded on each year, by land class	Regional Planning Data Base	Annual		xxx
		●Land use shift	Number of acres or percentage of "open space" used per year	Acres intruded on each year, by land class	County Agricultural Department Regional Planning Data Base	Annual	xxx	
		●Premature development	Number of acres developed in subregion or traffic zone, not in accord with general plan	Acres intruded on each year, by land class	Regional Planning Data Base	Annual	xxx	
		●New strip development	Along side of arterial roads	Acres intruded on each year	Regional Planning Data Base	Annual		xxx
		●Changes in housing density	Dwelling units/acre	Changes in intensity of land use		Annual		xxx
Recreational Resources (The Mitre Corp., under contract to CEQ, is producing an	All developed park land and accessible water area, including lakes, bays, ocean beaches	Developed park land	●Acres per 1000 people in county, unweighted ●Acres per 1000 people on basis of proposed	Total park acres inside the urban area; within 25 miles of and within 50 miles of the urban	U.S. Department of Interior, Bureau of Outdoor Recreation	Annual	xxx	

index of this type that will permit comparison of San Diego with 10 other urban areas.) Regional Planning has established a goal of 15 acres of park land per 1000 population, which would be factored into the expression of quality.	within 50 miles of the population centroid of the region, with a weighting applied to compensate for the distance from the population center to each park. Standards proposed for park land are same as those to be proposed by President's Council for Environmental Quality (CEQ).	● Within urban perimeter ● Within 25 miles of urban perimeter ● Beyond 25 and within 50 miles of urban perimeter	weighted values: ● Acres/1000 people with weight of 1 acre = 1 ● Acres/1000 people with weight of 1 acre = $\frac{2}{3}$ ● Acres/1000 people with weight of 1 acre = $\frac{1}{3}$	perimeter	California State Department of Parks County Park Department and Park and Recreation Departments of the Cities			
		Water-based recreation ● Accessible lake, bay, or lagoons	Acres of surface area: ● within 30 min. of population centroid = 1 ● in rest of county = $\frac{1}{2}$	Total surface area for water-based recreation	1970 Census Data, as updated by Regional Planning	Annual		xxx
		● Beaches	1 mi. beach within 30 min. of population centroid = 1 All other = $\frac{1}{2}$					xxx
Wild Life and Natural vegetation	This includes wildlife and its habitat, which is not part of developed park land category. Ecologically valuable niches are also included in this category even though they may be part of the "open space" indicator in the land use category.	● Acres of wild life habitat ● Number and total population of each species of ● Game and non-game animals, birds, fish, lizards, snakes, and other creatures	Total acres and ratio of wild life to acres, by species.	Number of acres Number of species and population of each	California Department of Fish and Game	Triannually	xxx	
				Game birds and animals taken during hunting season	California Department of Fish and Game	Annually	xxx	
		● Flora, number of species and abundance of each	Percentage change		IREM Inventory of Natural Resources	Triannually		xxx
		● Acres of especially valuable	Acres, and percentage of total,		IREM Inventory of Natural Re-	Triannually	cxx	

Table 2 – Continued

Quality Index Categories	Definition of Category	Proposed Quality Indicators	Method of Expressing Quality	Data To Be Collected	Data Source(s)	Frequency of Producing Index	Proposed for: Initial System	Proposed for: Ultimate System
		habitat—wetland and other	and sites lost		sources			
		●Threatened species of the region	Statement of present status	Status of troubled wild life	California Department of Fish and Game	Triannually		xxx
Natural resource depletion	Rate of use of renewable and nonrenewable resources of the region and impact on local economy of the unavailability of imported resources.	●Water table depth	●Number of months of reserve water in the ground	Depth of ground water		Deciannually		xxx
		●Prime agricultural land	●Percentage of total lost this year to development	Acres lost	County Department of Agriculture	Annual		xxx
		●Historical and archaeological sites	●Number threatened by development	Number of sites versus number threatened	IREM Inventory of Natural Resources	Triannually		xxx
		●Unique land forms						
		●Minerals—principally construction materials	●Percentage used this year ●Percentage for — years Supply remaining	Amount of material used by year for last 10 years	Local materials suppliers	Triannually		xxx
		●Imported water	●Percentage of total requirement that is imported	●Acre-feet of water imported versus acre-feet used	●Water Departments	Annual		xxx
		●Natural gas	●Impact of substituting liquid fuel for gas	●Ratio of liquid to gas fuel used	●Utility companies			
		●Others to be developed						

Table 3. Environmental Quality Index Categories and Indicators for Multimedia Categories

Quality Index Categories	Definition of Category	Proposed Quality Indicators	Method of Expressing Quality	Data To Be Collected	Data Source(s)	Frequency of Producing Index	Proposed for: Initial System	Proposed for: Ultimate System
Hazardous substances (including pesticides) Such as lead, mercury, arsenic, asbestos, zinc, DDT, polychlorinated biphenols More definitive data are expected to be available from the Monterey Basin Pilot Monitoring Programs. These affect air, water and land.	All toxic or otherwise known or suspected substances hazardous to man or the biosphere, which are capable of being monitored at the regional level, except those included in other indices (e.g., CO, particulate matter).	Amount to be sold or used	Pounds or tons	Quantity of hazardous materials sold for use in region	State agencies	Annually		xxx
		Levels in air samples	Micrograms or nanograms per cubic meter of air	Analysis of air samples	EPA's National Air Sampling Network and University of California at San Diego (Scripps)	Annually	xxx	
		Presence of hazardous substances in or on ● Fatty foods ● Water ● Human tissue ● Shell fish—bivalves ● Plankton ● Bottom sediment	Residual levels in ppm or ppb above or below standard or tolerable levels	Quantity of ● Pesticides (by type) applied by licensed applicators	State and County Departments of Agriculture	Annually		xxx
				● Level in water ● Level in shellfish, plankton, bottom sediment	Regional Water Quality Control Board	Annually	xxx	
				● Levels in fatty foods ● Levels in human tissue	Health Care Agency, Coroner's Office		xxx	
Solid Waste Disposal Includes land fill, incineration and, where practiced, ocean dumping of wastes.	This is a measure of the magnitude of the disposal problem and a measure of the efficacy of recycling programs plus an indication of hazards associated with waste disposal.	Amount of solid waste dumped	Pounds per year per person	Amount of solid waste processed	City and County Sanitation Departments	Annually	xxx	
		Recycled materials programs	Tons per year or percentage of total solid waste reclaimed	Amount of materials reclaimed		Annually	xxx	
		Ecological safety of solid fill programs	Adequacy of special fill areas for toxic substances	Evidence of leaching from solid fill sites		Annually		xxx

Table 4. Environmental Quality Index Categories and Indicators for Social-Aesthetic Categories

Quality Index Categories	Definition of Category	Proposed Quality Indicators	Method of Expressing Quality	Data To Be Collected	Data Source(s)	Frequency of Producing Index	Proposed for: Initial System	Proposed for: Ultimate System
Housing Adequacy Based on data that are to be produced on a triannual basis by HUD and Census Bureau commencing in 1973	A ranking of housing quality based on the estimated number and percentage of total dwelling units that meet each combination of criteria for adequate housing.	*Adequacy* ● Indoor plumbing—number of bathrooms/person ● Central or installed heating ● Electrical capacity ● Density and crowding—number of persons/room	Percentage of total units meeting each specified combination of requirements for adequate housing. ● Maximum density = 1.1 persons/room	On basis of samples of the Standard Metropolitan Statistical Areas, HUD to collect data on each indicator.	Joint Bureau of Census, HUD. Triannual survey to be started in 1973 for each of 60 SMSAs.	Triannually		xxx
		Availability ● Vacancy rate ● Number of families paying more than 25 percent net income for housing	● Vacant units/total units ● Number of families as a percentage of total families in the SMSA	● Vacancies ● Number of persons applying for housing subsides (per FHA)	Regional planning and local FHA housing survey	Annual	xxx	
An Alternate Approach is to use 1960 census data on "Substandard Housing," being updated for release in 1972	Substandard is unsafe, unsanitary, or dilapidated.	Substandard housing	Percentage of total units in San Diego County that are substandard	Data available from Census Bureau by December 1972 by county	Vol. 6 of 1970 Census Report, "Substandard Housing"	Deciannually		xxx
	Overcrowding of living space	Number of persons per dwelling unit	Percentage of dwelling units with more than 1.1 persons/room			Deciannually		xxx

Visual Pollution	A measurement of the visual impact of man on his environment	Billboards, unscreened junk yards, power lines, abandoned cars, rundown and abandoned structures, nonintegrated architectural complexes, littered open space, littered streets	Average number of pollution units per mile, based on walk-through of significant sample of streets of region	Pollution units, based on observations by trained person	San Diego Environmental Development Agency, Parks Department, or college students to conduct survey	Annual		xxx
Pests	Undesirable animals and insects that represent a health hazard or are a source of annoyance to the community	1. Actual number of rabies treatments administered	1. Number of treatments per 100,000 population per year	1. Number of cases	Health care agency	Annual		xxx
		2. Number of citizen complaints of mosquitoes, flies, roaches, rats, etc., that are verified	2. Verified complaints per 100,000 population per year	2. Verified complaints		Annual		xxx
		3. Number of stray animals impounded	3. Number of animals impounded per 100,000 population per year	3. Number of animals collected		Annual		xxx
Other Social/Economic Indicators	A set of indicators will be selected, using data currently available to the Environmental Development Agency.					Annual	xxx	

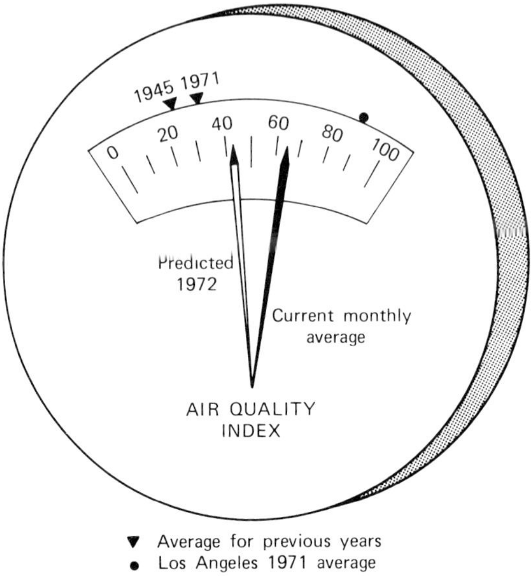

Figure 1. San Diego air quality index, ORAQI. An ORAQI reading of less than 20 is considered excellent. (Numbers are for demonstration only.)

Formats for Presentation

Recognizing that different uses of the indicator data requires some different formating of the information, some initial effort was spent in defining formats that would simplify a complex subject. In this regard, the proposed treatment of Air Quality is worth reviewing. The study recommended the use of ORAQI Index (Oak Ridge Air Quality Index), which integrates several physical measurements of air pollutants. It was felt that we could develop a display such as a gauge or meter, with which most people are familiar, that would indicate at a glance current and long-range trends, revealing "bad" or "good" by the direction of movement of the indicator. Such meters could be used in an IREM control room or on a weather program report on local television. An example of one such display appears in Figure 1.

It was also recommended that an annual environmental quality report be prepared for release either as part of a Board of Supervisors Annual Report on the "state of the county" or for release to news media.

Implementation Plan

Some 42 subindicators in the four major categories were recommended for development over time. Development of the implementation plan brought into focus questions of resources, availability, and related subjects. Based on the relative value of environmental information, the accessibility and cost of the data, the availability of IREM staff and resources to support development, and the urgency of need to establish communications, a set of initial indicators was proposed for a Phase II effort. This has been called the Initial Operating Capability (IOC) plan.

Phase II Activity[3]

Implementation

In October 1972 work was commenced to implement the IOC. Work has progressed to achieve the following objectives:

- Final determination of indices to be developed.
- Design of a data base for the time series of basic statistics to be collected for each area of environmental concern.
- Development of formal index methodology for each area.
- Development of a file of available statistical data.
- A report on the state of the environment of the region.

The first task resulted in reorientation of the indices to be developed. One of the principal changes was the addition of indices to describe the status of energy production and consumption in the region. It was determined that the availability of electrical energy from stationary sources may be a limiting factor in land use and a major contributor to pollutants in several of the environmental media. Consequently, energy was included in data bases and indices under development. Social and Aesthetic indices were determined to be of less immediate concern and their development was postponed.

Data base design has now been completed, and methodologies for formal index computation have been determined for the indices.

Environmental Quality Indices

Following is list of the EQI under development and for which data base design is complete. In most cases, community or subregional breakdowns of the data are possible. Where this is not currently possible, collecting agencies are being urged to add this criterion to their data collection practices.

Air Quality

- ORAQI Index—With norms for daily, monthly and annual averages.
- Comparative Day Indicator—Potential modification of ORAQI Index to include effects of wind, temperature, precipitation, and inversion layer.

- Total Oxidant Index—With daily norms for different months of the year.
- Episode Days Indicator—Number per month and year.
- Emissions Index—Tons per day for each category of pollutant and emitter (from Air Pollution Control District data).

Water Quality

- Bays and Lagoons Index—Percentage of surface area meeting quality standards.
- Fresh Surface Water Index—Percentage total volume of fresh water that meets standards for swimming and drinking.
- Ground Water Index—Percentage of total ground water meeting criteria of highest beneficial use planned—Regional Water Quality Control Board samples.
- Ocean Water—In shore ocean water near effluent outfalls by coliform, dissoved oxygen, secci disk measures; percentage of shoreline unsafe.
- Imported Water—An index of fresh water usage, showing past and projected consumption by various categories of users.

Noise

- Impacted Area Indices—Measurements of noise levels in decibels. A scale—dB(A) in and around airports, highways, and other designated impact areas. Percentage of people in subregion affected and number of acres affected.

Land Use

- Rate of Urbanization Index—Number and percentage of open space developed per year; number and percentage of acres in urban, urban fringe, etc., category.
- Land Use Shift Index—Number of acres converted to higher (or lower) density use, by type use.
- Recreational Use Index—Number of acres of developed recreation space per 1000 population (compatible with CEQ criteria).
- Coastal Plain Index—Number of acres in plain—percentage of acres filing for 80% level of development exception, etc.

Energy (Stationary Electric Power)

- Energy Production Index—Amount produced per year by type source.
- Energy Demand Index—Amount used and projected future use by category of user: industrial, commercial, residential, other (including export).
- Supply Cost by Stationary Source—Natural gas, fuel oil, nuclear, other.

- Estimate of Future Supply Costs—Same as above, including estimated reserve life.
- Pollution Contributions from Sources to Media—Estimated tons pollutant/Mega watt hour into air, water, noise, solid waste EQ media—effects of fuel alternatives (e.g., increased nuclear); possible constraining impact on Land Use of energy supply levels.

Solid Waste

- Solid Waste Index—Tons per day developed; cost per ton.
- Land Fill Indices—Numbers of acres of land filled per year—reclaimed. Acres in available sites.
- Recyclable Index—Percentage of total Solid Waste reclaimable. Percentage reclaimed.

In February 1973 a qualitative appraisal of the state of the environment was partially completed. This is to utilize data from the indicators just given as well as other statistical data available to yield a comparative picture of San Diego's environment and its outlook.

Notes and References

1. Don Macdonald and D. G. Malcolm, "An Approach to the Development of an Environmental Quality Information System for the San Diego Region," A Report to Project IREM, Research Analysis Corporation.
2. "Environmental Quality," Second Annual Report of the Council on Environmental Quality, August 1971.
3. The authors acknowledge the assistance and participation of L. E. Coate, Pat Bonner, John Hemphill, T. Deberry, and S. Stanik in this phase of the project.

Editors' Note: The final project report served as the basis for a State of the Environment Report for San Diego County, still pending publication through the county's Office of Environmental Management.

Indices for Policy Planning

ROY A. PAUL

The State of North Carolina is currently engaged in a process of planning and policy formulation for the future. We are seeking to raise the economic and social indicators of the state, primarily the per capita income. At the same time we wish to preserve the best aspects of a pleasant and livable environment. Although we have had social and economic indicators to use in such planning, we have not had adequate indicators of environmental quality. Therefore, during the past year state government has developed a comprehensive set of indicators and has acquired data on six primary indices.

Environmental Policy in North Carolina

Policy decisions in state government occur at different levels, just as in federal government or in substate regions. The legislature has the most broad powers, and policy is expressed in legislative action. The governor can set executive policies, make appointments, and recommend actions to the legislature. Certain boards and commissions have been established in the executive branch which function independently within given areas of responsibility. In each of these three levels of government, policies must be set and decisions made. In each of these three levels, formal or informal planning processes aid in the making of those decisions.

The level of concern in this chapter is that of the governor. He can promulgate policy formally by executive orders or policy statements, he can adopt a position in a speech, or he can express policy informally through his actions in making appointments to boards and commissions or recommending certain budgets and legislation. When these formal and informal statements of policy affect the environment, they may be labeled "environmental policy." We have

Mr. Paul is with the Office of State Plannning, North Carolina Department of Administration.

titled the formal process leading to environmental policy "planning for environmental quality."[1]

Governor Robert Scott, whose term of office expired on January 10, 1973, expressed a long-standing desire to help plan for the future of the State. In 1971 he established a North Carolina Council on State Goals and Policy, composed of 14 citizens of different backgrounds chosen from across the state. Chaired by the governor, the Council expresses the needs of citizens and identifies the kind of future they want in the form of recommended state goals; it evaluates the present activities of state government, identifies also areas of urgent public need, and informs the public of the priority problems facing the state. It is apparent that no single group could respond in depth in all areas of interest to the state. The field of environment is but one subject, whereas the Council's concern is the sum of all the concerns of government. The first annual report of the Council expressed goals in four areas: the economy, learning, health, and the environment.[2]

To aid the Council in the environmental area, an Interagency Task Force on Environmental and Land Use Planning was established in January 1972. The task force effects coordination among the many agencies of state government concerned with environment, provides guidance for environmental planning projects, and provides advice and policy recommendations to the Council on State Goals and Policy. The Interagency Task Force represents 21

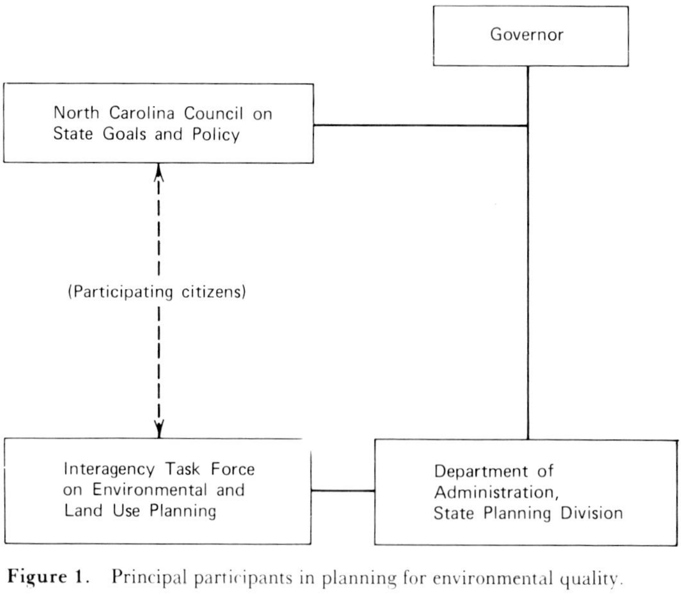

Figure 1. Principal participants in planning for environmental quality.

operating agencies or agencies of state-financed universities. It includes the air and water quality programs, assorted conservation programs, and research institutes of the major universities. Staff is provided by the Office of State Planning, within the Department of Administration[3] (see Figure 1).

The Environmental Indicators Project

The first job undertaken by the Interagency Task Force was the development of concepts of environmental indicators. First the staff assembled a collection of ideas about environmental indices from a variety of sources, including Oak Ridge National Laboratories, the President's Council on Environmental Quality, Mitre Corporation, the Environmental Protection Agency, and the National Sanitation Foundation. These concepts were distributed to the task force to illustrate the degree of current interest and the relevance of the subject. Second, a meeting of the task force was held in January 1972 to further motivate the program. We were greatly aided in this instance by Dr. William Thomas of the Oak Ridge National Laboratory. Third, each agency was encouraged to submit its own concepts of environmental indicators, using a prepared format to briefly outline the concept or idea. The response was enthusiastic. Within 60 days we had assembled more than 80 ideas for methods to measure environmental quality. Over the next four months these concepts were refined, revised, or expanded and were published in a progress report entitled *Planning for Environmental Quality, Phase I*. During this period, the staff conferred with agency representatives to further explain and to stimulate the creative process. It should be understood that the real process involved was negotiation, or an attempt to reach an agreement between the planning agency and the operating agency responsible for a given environmental program. This process represents the crux of our experience with the development of environmental indices.

It is important to understand our concepts of what environmental indices are and how they are supposed to be used in an environmental policy planning process. From the first, we attempted to allay fears about the indicators by clearly explaining how they were to be used. We expressed this as four applications:

1. *Environmental indicators provide a working definition of environmental quality*. Different agencies and different individuals have different concepts of what environmental quality is. This project gave them an opportunity to express their ideas for the first time in concrete terms.
2. *Environmental indices provide a simple format for communications*. During policy planning, the environmental indices were conceived as means for professionals on the task force to convey to the citizens on the governor's

council an overview of the "state of the environment." Using environmental indicators, this overview could be expressed in trend charts and maps illustrating the regions of the state exhibiting the worst deterioration. Other users of this information will be state agencies, legislators, citizens' groups, and the lay public. In this regard we are presently preparing an information brochure to be entitled "The State of the Environment in North Carolina."

3. *Environmental indicators provide a format for "operational" goals and objectives.* Probably the most important use of indicators is to provide a simple format for goals statements. For example, the state could set an operational goal of reaching a water quality index of 100 throughout the state by 1980. If such a goal were set, the agencies would know exactly what deadlines to aim for and what legislation they would need; in addition, they could estimate the budget required.

4. *Environmental indicators can be used to chart trends or changes in environmental quality.* Once the state has set operational goals, the indicators can serve in charting progress over the years to see whether those goals are met or whether changes in legislation or programs are required.

5. *An added function of indices will be regional planning and modeling.* We are proposing to determine the feasibility of modeling the relationship between policies and environmental indicators. We would project environmental trends in the Appalachian region of North Carolina under alternative population policies, land use policies, and environmental control policies.

Concepts Applied

In our discussions and negotiations we differentiated between indices and indicators. The term *indicator* represented any measurement of environmental quality or related aspect of the environment. We reserved the term *index* to apply to a single number that could reflect trends in a broad subject, such as water quality, and could be based on measurements of many different parameters.

The indicators, moreover, were to be based on data collected as a component of ongoing environmental programs. We have assumed that the information system is less than ideal but that it was necessary to utilize the objective information currently available rather than to wait for an ideal information system to materialize.

During the negotiations some professionals viewed the indicators differently from the project staff. They were concerned about trends in their programs and progress in carrying out certain environmental control strategy or presenting information on issues related to the environment. All this was relevant, but the focus differed from the view of staff, which was to develop a set of indicators on the environment itself from direct measurements of environmental parameters. This was resolved by categorizing indicators in such a way that all the concepts could be included (see Table 1).

During the first six months of the project, the initial 81 indicators were generated and documented. During the second six months, many data were collected and the concepts were negotiated further. It was found that much of the information was not in an easily utilized form. For example, it was decided that a water quality index should be based on data from 1700 water quality monitoring stations across the state. However, these data were recorded in a single set of books at the Office of Water and Air Resources. Thus one component of data collection involved the computerization of all the state's water quality records.

Table 1. List of Categories of Environmental Indicators

A. Indicators of Environmental Quality
   1. Quality of the air
   2. Quality of water in streams, lakes, estuaries, and aquifers
   3. Quality of drinking water
   4. Abundance of wildlife
   5. Quality of housing
   6. Amount of recreation, historic, cultural, scientific, and scenic areas
   7. Amount of state resources preserved for future generations
   8. Safety of food produced and consumed
   9. Amount of solid waste in the environment

B. Environmental Systems Indicators
   1. Water supply purification and distribution systems
   2. Sewage collection and disposal systems
   3. Solid waste collection and disposal systems

C. Environmental Policy Indicators
   1. Energy policy
   2. Land policy
   3. Lake and reservoir shorelands
   4. Flood plains
   5. Quality of forests
   6. Erosion and sedimentation
   7. Economics
   8. Population
   9. Artificial lake development
   10. Radiation
   11. Mining operations

Studies in the literature on environmental indicators tend to focus on the development of air and water quality indices, and the work involves a number of basic issues of interest to professionals and researchers alike. We believe that interdisciplinary negotiation on this subject has yielded a conceptual approach that is professionally sound and based on currently retrievable monitoring data. This concept is the "standard environmental index." The standard environmental index is based primarily on the work of two students employed by the state for three months in 1972: James McGinnis, an undergraduate of Stanford University, and Susan Wyatt, a graduate student in environmental science from the University of North Carolina.

The development of any index may be described in four steps requiring agreement by some professional body:

1. The parameters must be selected.
2. A method must be devised for translating measurements for different parameters into a common scale.
3. The importance, or weight, to be given different parameters must be decided.
4. A method must be chosen for aggregating data so as to arrive at an index number representative of a large geographic area for a given time period.

The following method appears to resolve the four issues raised by the use of indices in a manner satisfactory to professionals in North Carolina.

1. All measurements taken over a given geographic area are used in the calculation of the index, *if* a federal or state standard is applicable to the measurement. Since some measurements, such as biological oxygen demand (BOD) in a stream, are taken to analyze the nature of the pollution, no standard is directly applicable. Parameters with applicable standards, such as dissolved oxygen in water or particulates in air, are utilized.
2. All measurements are converted to common units by expressing them as a ratio to their respective standards. The common units, or index units, may be considered to be measures of the "percent of standard." The standard is placed in either the numerator or the denominator of the ratio such that an index unit of 1.00 represents the measurement exactly at standard and index units above 100 are worse than the standard; those below are better than standard.
3. Weighting and aggregation of data is accomplished by using both the mean and the "worst" values for all measurements in all samples taken from a given geographic area. The worst value may be approximated by the 95th percentile, to omit laboratory or sampling errors.

It will be noted that more weight is given to those measures that deviate most widely from the standard. This is quite evident for the "worst value," for it depends entirely on which measured value varied farthest from the standard. It is also true for the mean, because the index units may vary greatly on the upper side of 1.00, but index units that are less than one may only vary from 0 to 1.0, by the very method of calculation.

The principal advantage of the *standard index* is its reliance on official federal or state standards for determining the significance of measurements of environmental quality. Such standards are very acceptable among scientists, engineers, and administrators because they are based on current research and the judgment of experts well respected in their field. The standard index is a measure of variance from these standards.

Observations

From our experience in the development of environmental indicators we would like to offer certain observations and recommendations. The successful negotiation of a comprehensive set of indicators can be a long and complicated process. Under rather good conditions we completed the process from conceptualization through negotiation to data summarization in just 14 months. However, such a crash program is not always feasible, and two to three years would represent a more realistic program under differing circumstances. Moreover, we undertook to develop indices as a matter of necessity, because "official" indices were not available from other sources. Now, however, more knowledge and experience has been gained both locally and nationwide, and it may be wasteful for every state and every region in the country to develop its own concepts. Our recommendation is twofold:

1. *For local government.* Utilize wherever possible indices that are acceptable, regardless of their source, rather than spend too much time or money on the preliminary stages of the planning information system.
2. *For the federal government.* Develop as rapidly as possible a recommended set of environmental indices. It is desirable to have an "official" method for the utilization of environmental data; otherwise 50 states and thousands of local governments will have to duplicate the process to arrive at an acceptable methodology. Contrary to the opinion of researchers, these indices *do not* have to achieve optimum validity in a single number, only optimum officiality. The official methods of calculating indices should always be open to change and modification as new procedures and new methods become available.

The process of arriving at environmental indices is analogous to the negotiating process leading to environmental standards. Both are ostensibly based on extensive research on a myriad of physiological and sociological factors. But our studies are not sufficiently exacting to lead to a single methodology that can achieve total consensus. Decisions boil down to a matter of negotiation between the many agencies, bringing forth their professional interests and their special biases and prejudices. Thus the process of negotiating environmental indices for the planning of environmental policy represents in itself an environmental planning process, and the final product represents the nucleus of a formal environmental plan.

References

1. Roy A. Paul, "Environmental Planning: A Waif is Adopted," *Planning for Progress,* **5,** No. 2, Winter/Spring, 1972.
2. North Carolina Council on State Goals and Policy, *First Annual Report,* November 30, 1972, pp. 37–53.
3. Roy A. Paul, "Environmental Planning: The Toddler Crawls," *Planning for Progress*, vol. **5,** No. 4 (Fall 1972), p. 9.
4. Roy A. Paul, in cooperation with the Interagency Task Force on Environmental and Land Use Planning, *Planning for Environmental Quality, Phase I,* June 1972, 103 pp.

Implementation of Environmental Indices

WILLIAM A. THOMAS

The subject of environmental indices[1, 2] certainly is a fashionable one, but we should not pretend to be breaking new intellectual ground. Also, we should keep in mind that our having time to reflect on the general quality of life indicates that ours is rather high indeed; nonetheless, much remains to be accomplished.

One prefatory remark is in order. Although here we are specifically considering the social or cultural aspects of environment, biological indicators[3] are not included. That part of the "environment" also must be part of a comprehensive indicators system. Now, however, I want to discuss the implementation of environmental indices as they relate to the National Environmental Policy Act of 1969 (NEPA) and the suggestion of a specific approach suitable for more general purposes.

National Environmental Policy Act

To comply with NEPA, the president reports annually to Congress on the "status and condition of the major natural, man-made, or altered environ-

Mr. Thomas was the Leader of the Environmental Indices Group, Environmental Program, supported by the National Science Foundation RANN Program at Oak Ridge National Laboratory (operated by Union Carbide Corporation for the U.S. Atomic Energy Commission). Now with the American Bar Association in Chicago, Mr. Thomas also edited Indicators of Environmental Quality: An Overview, Plenum Press, New York, 1972.

mental classes of the nation" and the "current and foreseeable trends in the quality . . . of such environments."[5] The Council on Environmental Quality is responsible for preparing this report[6] and, therefore, with the task of devising or assembling the required environmental yardsticks.

Now that sufficient momentum exists toward developing reliable indices for various environmental components,[7] it is time to apply them to another provision in NEPA, the well-publicized environmental impact statements that are included in all "proposals for legislation and other major federal actions significantly affecting the quality of the human environment."[8] This provision requires a consideration by responsible officials of whether a federal action should be initiated or permitted to proceed. This involves assessment of (1) the current environmental quality in the affected region or sphere of influence, (2) the contemplated environmental quality after completion of the project, and (3) the comparison of benefits to be gained with detrimental influences in light of all reasonable alternatives. Simply stated, what we need to complete this process satisfactorily are scalar values—that is, indices.

Before NEPA it often was sufficient to argue, "Yes, this new facility will increase air pollution, but it also will increase the payroll by x million dollars." Promoters of the project emphasized the amount of money to accrue to the region, and this value in fact became the indicator of whether the project should be completed. The detrimental factors could not be predicted as accurately because they were so nebulous. The issue centers on trading some degree of environmental quality for some degree of economic, social, or political gain that the project is to produce. It is nearly an overwhelming assignment to perform this assessment in a scientifically defensible fashion without some specific estimates on the magnitude of the losses as well as of the gains.

Although the Council on Environmental Quality is making headway in developing environmental indices to meet its NEPA responsibilities,[9] other federal agencies, with the possible exception of the Environmental Protection Agency,[10] have not been as responsive.

Implementation of Indices

I would like to propose a way for governmental or private organizations to develop environmental indicators. The increase in omnibus environmental legislation and "little NEPAS" on the state level and the frequent delegation to the states by Congress of responsibility for environmental quality[11] make such a project particularly appropriate for the states. Similarly, the current trend toward approaching regional problems through interstate compacts should encourage the needed regional outlook.

Governmental units can implement a system of environmental indices by convening a task force composed of one individual from each subdivision concerned in any way with environmental affairs. Each representative could thoroughly review the statutory and regulatory responsibilities of his agency and propose to the group various indices to measure the quality of each environmental component for which his agency is responsible. After discussions within the task force to ensure that areas of overlapping authority are handled cooperatively, a network of indices would be established. One of the problems most likely to be encountered is the lack of data suitable for the desired index.[12] This realization itself is an accomplishment if it leads to modification of the monitoring systems. Assuming that an indices project is established, and there are very few scientific or political reasons for not having one, let us examine some of the advantages.

Representatives of the agencies would get to know one another, perhaps for the first time. Citizens and governmental officials frequently encourage greater communication among programs, and this is one concrete way of achieving it, as long as there is strong central support for the project.

A coordinated effort would provide an excellent opportunity to involve universities and other academic institutions in the practical aspects of public administration. The organizational unit involved need not be governmental; it also could be corporation or other entity. Public pressure and the demands of state and federal statutes and regulations make it imperative that businesses consider their environmental problems holistically rather than in the traditional piecemeal fashion. A prime incentive for them to do so is the Occupational Safety and Health Act of 1970.[13]

If corporations took the initiative in self-assessment and self-regulation, the present movement toward more governmental controls might ease somewhat. Since indices would assist in corporate decision making, many corporations might be interested because they would be the direct beneficiaries.

Indices also help the decision makers, who must report to someone, explain why they made certain decisions. There never is enough money to meet all the demands, and we all know that the public or private administrator's job is to allocate it for the best overall advantage, over both the short and, if possible, long terms. Environmental indices are designed to assist in these decisions and to communicate to others the rationale behind setting certain priorities. They provide a measure of performance.

An example process might be as follows. The regulatory agency—say, the one responsible for air quality—annually proposes to the elected officials its next year's budget. Because of various pressures, real and imagined, these officials are open to public criticism if they do not increase the budget of the agencies charged with environmental quality, even though available funds already are in short supply. One approach would be to ask the representative of the agency what percentage increase he received in the budget last year, what percentage increase he managed to achieve in air quality, and if by these quantitative measures he thought his performance justified additional money. After some discussion, the officials could ask for a prediction of the increase in air quality to be achieved with the requested increase in funds. At the next year's fiscal hearing, these figures and the agency's actual performance as measured by the air quality index would form part of the basis for assessing the return on money expended by the public.

Without doubt, there is going to be resistance from several quarters to any proposal like this one,[14] but interested citizens can bring about this change by calculating their own set of indices[15] and comparing the trends through time with budgetary outlays or other considerations, such as zoning changes and shifts in the types of dominant industries. If the responsible officials respond that the citizens do not have adequate data or adequate techniques to calculate realistic indices, they will find themselves in the position of having to cooperate with the citizens by using indices themselves[16] or by making the data available.

Assorted Matters

Three other points merit consideration. Regulatory agencies collect a plethora of environmental data from a variety of sites for a diverse array of environmental components or pollutants.[17] Environmental indices would efficiently reduce these raw data to terms that are useful to scientists, public officials, and other citizens. As Bowen recently stated, "Analysts and other numbergetters will have to abandon their traditional independent stances and start mixing it with the number-users. The conflict may be a bruising one, but the cause is just."[18] After all, what we seek is a more habitable planet, not just descriptors of its present condition.

Environmental indices often are compared with economic indicators, as in the expression "Environmental Dow Jones." In fact, however, the environmental ones exceed the economic ones in utility because they allow comparisons among geographic areas and show trends through time, as well.[19] On establishment of acceptable indices, we will need a standardized information outlet similar to that now provided for economic indicators.[20] Not until then will we satisfy the demand for reliable, comprehensive, and interpretable environmental data.[21]

Because developers of regional models often are also interested in using indices, I want to point out a fundamental difference in approach between the two. Indices are developed best on the national level to ensure wide applicability and standardized data bases and may be modified to meet local or regional circumstances when necessary. Models, on the other hand, often evolve from smaller to larger geographical areas. As a result, the indices desired as inputs, outputs, or checks on a model may not be readily available throughout the model's growth, thus limiting its usefulness.

Conclusion

In closing, I would like to respond to a rhetorical question that we all hear frequently: If we can put man on the moon, why can't we clean up the environment? The answer is relatively simple. To get from earth to the moon we knew three of the essential facts. We knew where we were, we knew where we needed to go, and we knew that rockets would be used, but conceptually the problem was a rather simple one. Cleaning up the environment is not only a technologically bewildering problem but also a politically controversial one. More important, it is difficult conceptually because we do not all agree on what we mean by environment, let alone know where we are on any universally accepted scale of environmental quality. Moreover, there is no uniformity of opinion regarding where we should proceed along that scale if we had one in the first place. Perhaps the amount of money we are willing to spend will set the upper limit.[22] It is clear that significantly improving the quality of our environment requires a national commitment at least as firm as the one that put man on the moon. Although environmental indices many not tell us how to get from one point to another, I firmly believe that they can describe where we are and where we should be going.

Notes and References

1. There is some confusion about the definitions of "indicator" and "index," and the terms often are used interchangeably. A common distinction is that "indicator" is a measure that reflects the presence or condition of a fundamental variable and that "index" is a composite value for an environmental feature for which more than one variable must be integrated.
2. Many organizations have expressed their support for development of environmental indices:

American Chemical Society, *Cleaning Our Environment. The Chemical Basis for Action,* American Chemical Society, Washington, D.C., 1969.

Council on Environmental Quality, *Environmental Quality,* The First Annual Report of the Council on Environmental Quality. Government Printing Office, Washington, D.C., 1970. Subsequent CEQ annual reports also state the need for additional environmental indices.

Environmental Pollution Panel, President's Science Advisory Committee, *Restoring the Quality of Our Environment.* Government Printing Office, Washington, D.C., 1965.

Environmental Studies Division, Office of Research and Monitoring, Environmental Protection Agency, *Quality of Life Indicators.* Government Printing Office, Washington, D.C., 1972.

Environmental Study Group, Environmental Studies Board, *Institutions for Effective Management of the Environment,* Part I, National Academy of Sciences–National Academy of Engineering, Washington, D.C., 1970.

Brown, R. M., et al. (National Sanitation Foundation), "A Water Quality Index—Crashing the Psychological Barrier," in W. A. Thomas, ed., *Indicators of Environmental Quality,* Plenum Press, New York, 1972, pp. 173–182.

Kimball, T. L. (National Wildlife Federation), "Why Environmental Indices?" in W. A. Thomas, ed., *Indicators of Environmental Quality,* Plenum Press, New York, pp. 7–14, 1972.

Panel on Systematics and Taxonomy, Federal Council for Science and Technology, *Systematic Biology. A Survey of Federal Programs and Needs,* Office of Science and Technology, Washington, D.C., 1969.

Scientific Committee on Problems of the Environment, International Council of Scientific Unions, *Global Environmental Monitoring,* International Council of Scientific Unions, Stockholm, 1971.

3. An introduction to the literature on this subject is provided elsewhere: W. A. Thomas, G. Goldstein, and W. H. Wilcox, compilers, *Biological Indicators of Environmental Quality: A Bibliography of Abstracts,* Ann Arbor Science Publishers, Ann Arbor, 254 pp., 1973.
4. 42 U.S. Code Section 4321 et seq., 1970.
5. Idem at Section 4341.
6. Idem at Section 4344. Executive Order 11514 expands on these responsibilities.
7. W. A. Thomas, ed., *Indicators of Environmental Quality,* Plenum Press, New York, 275 pp., 1972.
8. 42 U.S. Code Section 4332, 1970.
9. Executive Order 11514 Section 3(e) specifically states that the Council on Environmental Quality (CEQ) shall promote the development and use of indices to determine the effectiveness of programs designed to enhance environmental quality. The Third Annual Report of the Council on Environmental Quality (Government Printing Office, Washington, D.C., 1972) describes work toward this goal.

Russell E. Train, Chairman of CEQ, stated, "I am convinced that we not only *can* develop such indices, but that we *must* if the level of environmental policy and planning is to be improved." (Emphasis in original.) Editorial: "The Quest for Environmental Indices," *Science,* **178,** 121 (1972).

10. Environmental Studies Division, Office of Research and Monitoring, Environmental Protection Agency, *Quality of Life Indicators.* Government Printing Office, Washington, D.C. 83 pp., 1972.
11. For example, the Clean Air Act, 42 U.S.Code Section 1857 et seq., 1970, and the Federal Water Pollution Control Act, 33 U.S.Code Section 1151 et seq., 1970. There is a possibility of federal preemption of some environmental matters, such as setting radiological standards in the vicinity of nuclear power plants. *Northern States Power Co.* vs *Minnesota,* 447 F.2d 1143 (8th Cir. 1971). As the dissenting

opinion in this case makes clear, the U.S. Supreme Court has consistently recognized the right of states to protect the health and safety of their citizens unless Congress specifically intended to preempt.

12. The President's Commission on Federal Statistics presumably is trying to organize the data collection networks into bodies that achieve unified ends, to put a stop to the present procedure of merely compiling data without comprehensible formats or summaries. A. Etzioni, Editorial: "Fact-Crazy, Theory-Shy." *Science,* **170**, 391 1970.

13. 29 U.S.Code Section et seq. (1970).

14. If environmental quality is deteriorating, the responsible officials are unlikely to cooperate in educating the public about ways to measure the degree of deterioration. The data collected usually are available for public inspection but often not in understandable or usable form. It is also possible that officials might devise their own indices in such a manner that an overly favorable picture is presented to the public. Any new tool can be misused, but I believe that open communication of indices is preferable to selected individuals making decisions based upon preselected information, especially if the public can check the veracity of the indices.

15. This was a prime objective in publishing one air quality index. W. A. Thomas, L. R. Babcock, and W. D. Shults, *Oak Ridge Air Quality Index,* Publication ORNL-NSF-EP-8, Oak Ridge National Laboratory, Oak Ridge, Tenn. 10 pp., 1971.

16. In 1970 during the second session of the 91st Congress, bills were introduced in the House of Representatives (HR 17907) and the Senate (S. 3556) that would provide for the calculation of an Environmental Quality Index to be made available regularly. Neither was reported out of committee (Interstate and Foreign Commerce Committee in the House and Commerce Committee in the Senate).

17. The scope of the data collection effort is broad indeed: Smithsonian Institution, *National and International Environmental Monitoring Activities. A Directory.* Washington, D. C. 272 pp., 1970.

18. D. H. M. Bowen, Editorial: "Playing the Environmental Numbers Game." *Environ. Sci. Technol.,* 6,7 (1972).

19. Environmental indices are not dependent on well-established political units, thus can be calculated for very small areas, permitting a greater variety of comparisons than is possible with economic indicators.

20. For example, "Economic Indicators" is a booklet prepared monthly for the Joint Economic Committee of Congress by the Council of Economic Advisers and distributed widely throughout the nation.

21. The principal conclusion of a study prepared for Congress by the National Science Board of the National Science Foundation (*Environmental Science, Challenge for the Seventies,* 50 pp., 1971) summarizes the problem:

Environmental science, today, is unable to match the needs of society for definitive information, predictive capability, and the analysis of environmental systems as systems. Because existing data and current theoretical models are inadequate, environmental science remains unable in virtually all areas of application to offer more than qualitative interpretations or suggestions of environmental change that may occur in response to specific actions.

22. The amount of money that people are willing to spend to improve environmental quality is itself an indicator of concern for the environment. The latest survey results are encouraging but still not fully satisfying. T. L. Kimball, "What America Really Thinks About Pollution Cleanup (A Special Gallup Survey)." *Nat. Wildlife,* **10** (3) 18–19 (1972).

Comprehensive Indicator: Quality of Life

ROBERT C. LIVINGSTON

The "quality of life" (QOL) concept has received increasing attention as the focal point of converging economic, social, and environmental considerations. Serious attempts are being made to develop the concept into a useful tool for decision makers in the public and private sectors.

Perspectives of Different Disciplines

Attempts to measure societal conditions have been undertaken from the standpoint of several major disciplines, including economics, sociology, psychology, and the environmental sciences. Each has its own understanding of how values and ideals should be defined and manifested in the laws, norms, and sanctions of society. The crucial concern regarding quality of life cannot be adequately evaluated without an understanding of these various value perspectives.

The Economic Perspectives

For economists in general, methodology has emphasized quantity as opposed to quality. The stress has been on technical analysis, concentrating on input–output studies, econometrics, operations research, game theory, and linear or mathematical programming. Until recently economics has paid little attention to the relationship of such concepts as production and distribution, goods and services, commodities and performances, to the human actors who control them and who, in part, are controlled by them.

Mr. Livingston is with the Environmental Protection Agency, Office of Research and Development, Environmental Studies Division.

The Social Perspective

Policy analysts and decision makers often forget that their backgrounds of race, ethnicity, life style, and so on—their respective value bases—greatly influence their "scientific" approach to the development of indicators that will measure the quality of life.

To generalize, the sociologist can be said to approach the quality of life concept from a group, institutional, or societal perspective; the psychologist approaches it from the personal or individual standpoint.

In developing such social indicators, the sociologist must recognize that society is a community of individual and collective meanings to be identified and quantified. Today, even the term "social indicator" is not fully defined, conceptually or theoretically.

The Psychological Perspective

In *The Human Meaning of Social Change*, Campbell and Converse devote a chapter to "Aspiration, Satisfaction, and Fullfillment." Certainly, these terms convey the full force of the meaning of "quality" for modern individual's life.

One of the difficulties in discussing the quality of life is that little is known about the relationship between attitudes and behavior. The discipline of psychology addresses this relationship.

Abraham Maslow approaches the perspective of *individual* needs and values with a "needs hierarchy" of five levels; in ascending order, these are physiological (or survival); safety/security; social, ego; self-fulfillment (or self-actualization). A more recent theory is the eight-level open-ended theory of Clare Graves, who sees turmoil as caused by the transition in moving from one "need" level to another). Pervading most people's quest for their vision of "quality of life" is a need for significance, continuing growth and personal satisfaction and fulfillment, no longer just survival.

The Environmental Perspective

The environmental sciences attempt to integrate the knowledge of the physical sciences with the perspectives of economics, sociology, and psychology by relating the environment and the impact of man-made changes to the quality of life.

A number of social concerns can be identified when approaching the environment from the perspective of quality of life: the continued support of man by the ecological systems as we know them today; the availability of suitable land, particularly for agricultural and recreational use; and an adequate supply of air and water of suitable quality to support all forms of life. Included in these categories are wildlife, areas of natural beauty, and recreation areas. It is difficult, given the present state of the art, to interpret these environmental indicators in terms of human welfare.

Certain indicators of quality of life are a possible beginning point in determining environmental quality in terms of human welfare because they relate to health; for example, the extent of air pollution and the number or percentage of persons experiencing it, and persons living within certain radii of polluted water bodies.

Of vital interest to all concerned with the environment and the preservation of life, is the linkage between certain environmental factors and the factors relating to man's sense of what represents for him quality of life.

Perspectives of Different Life Styles

Each life style grouping, whether made distinctive by class, age, interest, economic level, education, or a combination thereof, tends to define "quality of life" in unique and characteristic terms. Each has its own value base. Such differences are heightened by demographic and geographic considerations. Add to these life style considerations the pluralism of American society, and the challenge of creating acceptable quality of life measures becomes readily apparent.

Measurement of Quality of Life

No one has yet figured out how to measure QOL. One of the difficulties in measuring it as opposed to economic or production factors, is that there are no "social dollars" flowing throughout the system which can be counted at critical points. The social "system" has characteristics vastly different from those of the economic system. There are no social costs, prices, or incomes to which most social variables can be converted, given the current state of the art; consequently a tradeoff is difficult if not impossible to make.

In considering measurement of quality of life, one must keep in mind that information and evaluation requirements will change through time as values change and knowledge expands.

The subjective and objective conditions of quality of life measures must be recognized and dealt with. By definition, objective conditions are empirical and reproducible. Heads can be counted or absolute scales can be applied. Census data and economic indicators deal with the objective. Quality of life indicators, on the other hand, may go beyond this to include the subjective. Subjective conditions or factors are not reproducible with certainty. They must measure feelings and attitudes, indicating the conditions and conditioning or "states of mind" of society and its citizens.

Quantitative methods can be used to determine the normative character of such information. However, comparison of an attitude or value held by one person with respect to a different attitude or value held by someone else may not be valid. Quality of life could be the most valuable tool for the planner and decision maker yet devised.

The process of planning should be seen as future oriented; therefore, the planner has to have an insight into the variables affected by the implementation of a policy. From understanding the present holistically, the planner can see that he has three distinct and sometimes totally different paths to follow:

1. Staying with the present continuum on which current trends will evolve and shape his environment, planning is direct and goal-oriented. However, the ramifications on all other systems are not taken into account and planning is rudimentary, striving only to achieve the planner's unidirectional goal. The confidence limits or bounds within which his planning is oriented are not complete or wide enough, and they are transitory. This planner suffers from "tunnel vision," for his scope is narrow and he is satisfied when he reaches his goal. The long-run outcome may be a total waste of time, effort, and money.

2. The holistic approach projects the end result into the future in an effort to assess those ramifications for which the product would hinder, harm, or in some way have a detrimental effect. This planner's scope is broader than the first because he tries to understand what effect his desired goal will have on the entire system. This may seem to be an impossible task, for one cannot understand anything holistically; however, one can try to be as complete and thorough as possible. The planner must guide and constantly monitor the situation he is in relative to where he wants to go, keeping in mind the long-range effects. This is by no means simple, for without knowing the future one cannot tell what consequences his actions will have. The best he can do is try to forecast certain aspects about the future with which his objective may conflict and work out a plausible alternative. He may plan with given solutions to specific outcomes that he may forecast, thereby making the policy-planning process easier for the future. This job—the job of the comprehensive planner—is the basic building block for the future and the serious planner will follow this methodology.

3. The third continuum consists of nondirected, non-goal-oriented future planning activities. Change is brought about for its own sake, sometimes with little or no justification. The goal of the policy maker may be known; however, means may not be available to him at the time. For this reason the policy maker almost never has a chance to mandate a plan that will set him on a course to prevent the recurrence of similar crises in the future. Consequently, he is forced to make up a series of short-run, extrapolatory system patches, which he thinks will help to achieve his goal. This is often a stimulus–response

approach to a crisis situation which demands immediate results with little time to evaluate long-run occurrences.

In each of the foregoing cases, a goal is aimed at and in most instances reached. The differences lie in the methodology used and the detrimental consequences for the future.

Since the planner's purpose is to develop a planning methodology that prevents conditions that would significantly destroy the "quality of life" as perceived by his and future generations, the development of a QOL matrix would seem to be an obviously useful tool in planning for the future. The difficulty lies in the defining of QOL, for it has never passed the definitional stage of development.

Indicators are subject to diminishing returns in that each indicator's contribution in helping to determine the level of living has—in the long run—a turning or saturation point. Once this point is passed, the contribution is no longer favorable to the individual; in fact, rather than raising his level of living, a rising indicator tends to lower it.

Generating and Weighting Factors

From May through August, 1972, members of the Environmental Studies Division (ESD), Office of Research of the Environmental Protection Agency, developed a list of factors relevant to the concept of "quality of life." Because "authorities" on quality of life had never met at one time to analyze the concept and its applicability and usefulness to the decision maker and the policy implementor, ESD arranged such a QOL conference, on August 29–31, 1972, to review, analyze, and (hopefully) accept the listed factors. On August 30, the attendees were divided into five groups of 30 and presented with the list developed by ESD (see Table 1).

The groups were told to review, examine, expand, and/or delete the items in the list under five categories in one hour. If one group felt that a factor did not belong in its category but in another, it was to write down the factor to be presented to the group dealing with the other component. By this method, the members of the conference generated a list of QOL factors as depicted in Table 2. The five major categories of the original list were cut into three, combining political and social into one category and man-made and natural environments into "environment."

It would of course be presumptuous, and scientifically untenable, to attach any great significance to this factor list. There are numerous ambiguities in the definitions of induvidual factors, and there is even duplication (the factor "health" appears under two components). Furthermore, there is no uniformity in the scope of individual factors: "Democratic process" and "solid waste"

Table 1. Component/Factor List

Natural Environment	*Man-made Environment*
Overall pollution—the state of the environment as it relates to all forms of pollution.	Leisure facilities—the availability of resources (nonpersonal) for use in leisure activities.
Air pollution—the amount of foreign and harmful substances in the air from all possible sources.	Work environment—the working conditions consisting of location, physical surroundings, personal relationships, and job status.
Water pollution—the amount of foreign and harmful substances in the water from all possible sources.	Housing—the availability of adequate housing facilities.
Noise pollution—the production of sounds at a level which can be considered annoying and possibly harmful.	Technology—the state of our applied scientific capability as it relates to everyday life.
Aesthetics—the beautiful and/or pleasing aspects of the natural environment.	Aesthetics—the beautiful and/or pleasing aspects of the physical man-made environment
Land use planning—the well-organized, planned, and presumably beneficial use of the land.	Transportation—a combination of services and facilities provides for the mobility of the population.
Ecosystem—the flourishing of animal and plant life in support of the natural balance.	Material quality—the capability and reliability of manufactured products.
Radiation—the release of radioactive materials into the environment.	Utilities—the capabilities, effectiveness, efficiency, and reliability of the utility systems (water, gas, electricity) both public and private.
Pesticide/chemical—the use of pesticides and chemicals and their subsequent release into the environment.	Physical structures—the availability and/or condition of offices, factories, public buildings, etc.
Soil quality—The ability of the soil to support the needs of man as well as animal and plant life.	
Solid waste—the problems engendered by the requirements to effectively and permanently dispose of the residual matter remaining after the initial use and/or consumption of society's products.	

Table 1.—(Continued)

Social Environment

Education—the knowledge, skills, character, etc., of the population and the process of obtaining and developing these by formalized schooling and/or training.

Family structure—the arrangement or interrelation of the family unit with emphasis on cohesiveness and indivisibility.

Status—one's position, rank, or standing within the societal structure.

Culture—the concepts, habits, skills, arts, instruments, institutions, etc., of a given people in a given period.

Privacy—the ability to withdraw from public view or company; seclusion or secrecy in a given situation; and limitation of use of information concerning a person's background, status, etc.

Safety—freedom from danger, injury, or damage.

Social stability—degree of cohesiveness within a societal structure and between societies; war/peace; absence of riots and other disturbances; law and order.

Personal skills—degree of ability, proficiency, or expertise in a given area or areas.

Equality—state or instance of being equal in political, economic, or social rights.

Political Environment

Opportunity structure—a combination of circumstances favorable for and/or offering a good chance or occasion for individual advancement in relationship to ability.

Information media—the freedom, responsibility, and reliability of information sources that are public and/or private.

Democratic process—the acceptance and practice of the principles of equality of rights, opportunity, and treatment.

Civil liberties—rights guaranteed to the individual by law, such as thinking, speaking, and acting as one likes without interference or restraint except in the interests of the public welfare.

Justice—the use of authority and power to uphold what is right, just, or lawful.

Economic Environment

Living costs—that portion of total income required to satisfy basic human needs such as food, clothing, and shelter.

Income—the money or other gain periodically received by an individual, corporation, etc., for labor, services, or from property investments, operations, etc.

Income distribution—the stratification of income in relation to various sections of society.

Table. 1—(Continued)

Social Environment	*Economic Environment*
Choices in life—the chance, right, or power to make fundamental choices in life usually by the free exercise of one's judgment and the use of one's resources.	Consumption—the utilization of economic goods or services in the satisfaction or wants or in the process of production.
Community—a group of people living in the same area living under the same culture and norms and having interests in common.	Economic security—the relative degree to which one is satisfied; secure in his economic state with relation to his needs and requirements for his level of aspiration.
Health—physical and mental well-being; ability to detect, treat, and control specific causes and characteristic symptoms of illness and ailments. Quality and quantity of services provided in order to maintain mental and physical growth and development.	Economic growth—the change over the previous year in total production of goods, total economic activity, total consumption, or total productive capacity.
	Labor—the quantity and/or quality of the labor force in terms of productive capacity, specialized skills, and diversity.
	Capital—the accumulated and concentrated wealth used to finance the creation, operation, and expansion of economic activity.
	Investment—the outlay of land, labor, and/or capital to the future production of goods and services.
	Public spending—the amount of current and consumable resources that are channeled through the public sector for either direct redistribution or final consumption by public agencies.

Table 2. Revised Component/Factor List

Political/Social

Democratic process—the system by which the public governs itself in accordance with the principles of equality of rights, opportunity, and treatment.

Public participation—the ability of citizens to be heard and to interrogate the decision makers.

National security—the provision and dispersal of military forces and other resources in a manner determined by the democratic process and reasonable public participation.

Education—the knowledge, skills, etc., of the population and the process of obtaining and developing these by schooling, training and other means.

Status—one's position, rank, or standing within the societal structure.

Culture—the concepts, beliefs, habits, skills, arts, instruments, institutions, etc., of a given people in a given period.

Privacy—the ability to withdraw from public view or company; seclusion or secrecy in a given situation; and limitation of use of information concerning a person's background, status, etc.

Safety—freedom from danger, injury, or damage, physical or psychological.

Economic

Accumulated assets—the assets owned by an individual in the form of savings, property, insurance, etc.

Essential living costs—that portion of total income required to satisfy basic human needs such as food, clothing, shelter, etc.

Income—the money or other gain received by an individual in the form of wages, salaries, tips, dividends, rent, interest, etc.

Income distribution—the distribution of incomes in relation to various sections of society.

Consumption—the utilization of economic goods or services in the satisfaction of wants.

Economic security—the degree to which one can maintain his economic state in relation to his needs.

Discretionary income—the money remaining from the individual's income after essential living costs have been deducted.

Capital—the accumulated and concentrated wealth used to finance the creation, operation, and expansion of economic activity.

Investment—the outlay of land, labor, and/or capital to the future production of goods and services.

Table 2.—Continued

Political/Social

Social stability—degree of cohesiveness within a societal structure and between societies; war/peace; absence of riots and other disturbances; quality of law and order.

Personal skills—degree of ability, proficiency or expertise in a given area or areas.

Equality—state of instance of being equal in political, economic, legal, or social rights and opportunities.

Choices in Life—the chance, right, or power to make fundamental choices in life usually by the free exercise of one's judgment and the use of one's resources.

Health—physical and mental well-being. Ability to detect, treat, and control specific causes and characteristic symptoms of illness and ailments. Quality and quantity of services provided in order to maintain mental and physical growth and development.

Primary social relationships—a personal, informal relationships of the individual to others within families and other circles of intimates.

Secondary Social Relationships—impersonal, formal relations of the individual to others in work and other community environments.

Use of free time—amount of free time alone or in voluntary associations (e.g., church clubs) and variety of activities engaged in.

Economic

Rate of population growth—the rate at which the total population is increasing.

Economic opportunity—circumstances favorable to improvement of an individual's material welfare in relation to his ability.

Economic development—capacity to utilize resources, natural and human, and capital stock to achieve greater and higher quality output.

Enviroment

Air and Water Categories (measures quality and quantity)
- Air
- Water
- Noise
- Climate/Weather
- Hazardous substances

Land-Related Categories (measures land shifts and quantities)
- Land use
- Recreation resources
- Ecosystem
- Natural resource depletion
- solid waste

Man-Made Categories (measures availability, accessibility, aesthetics and design, product quality, and ecological linkage)
- Employment
- Housing
- Transportation
- Cultural and spiritual
- Commercial and service
- Utilities
- Education
- Leisure

have equal stature in the list. Structurally, the final list consists of some 47 factors grouped under three component headings:

- Economic
- Political/Social
- Environment

That a wide variety of factors were grouped under one super-component,

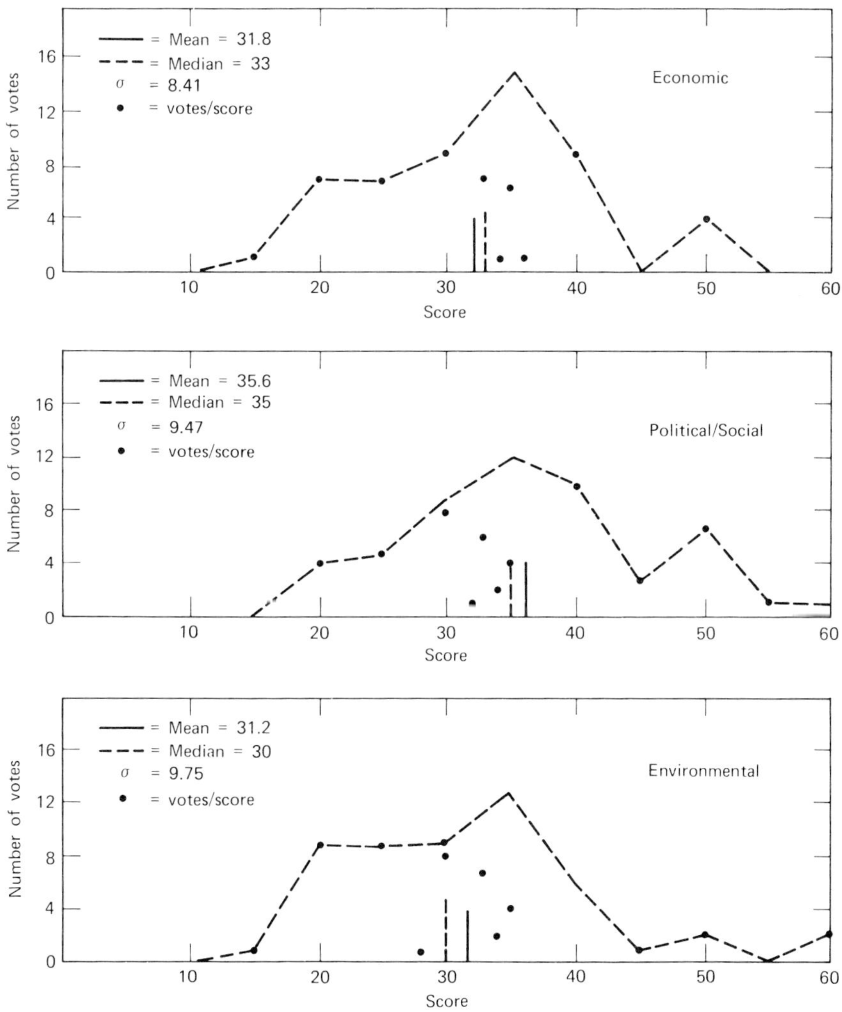

Figure 1. Component distributions and statistics.

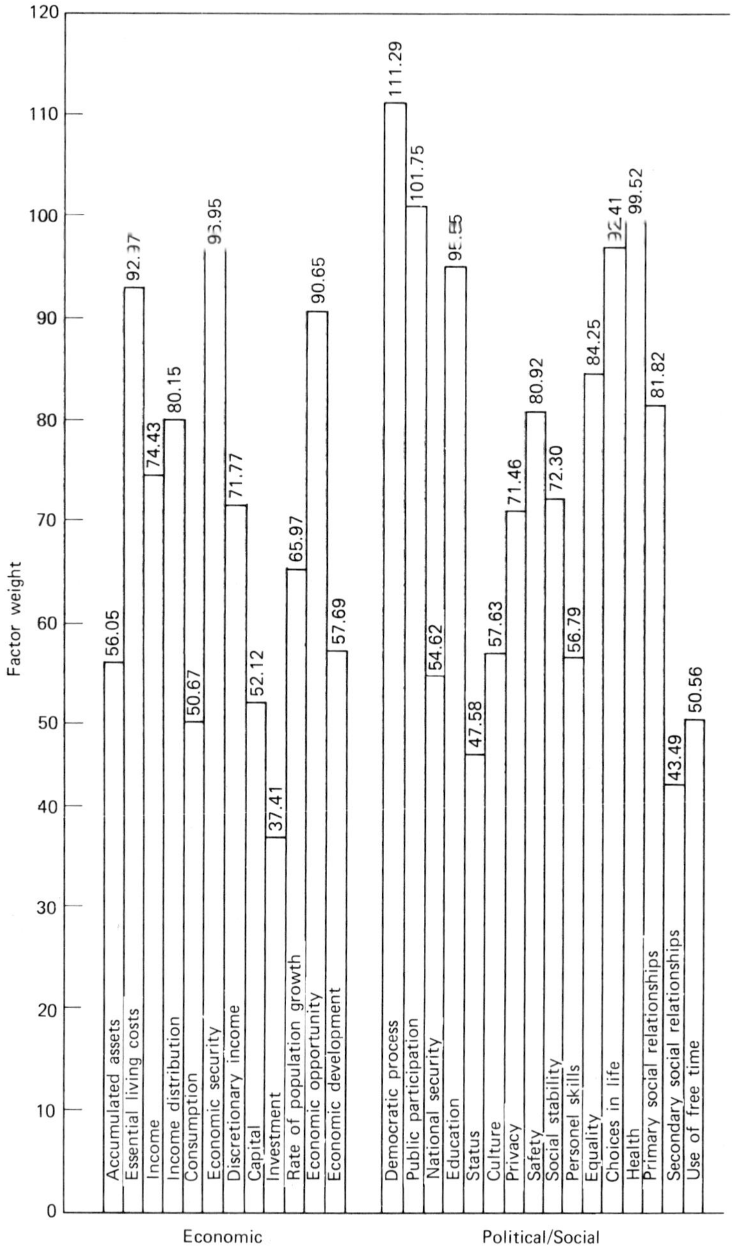
Factor weight
120
110
100
90
80
70
60
50
40
30
20
10
0
Economic
Political/Social
Accumulated assets
56.05
Essential living costs
92.37
Income
74.43
Income distribution
80.15
Consumption
50.67
Economic security
Discretionary income
71.77
Capital
52.12
Investment
37.41
Rate of population growth
65.97
Economic opportunity
90.65
Economic development
57.69
Democratic process
111.29
Public participation
101.75
National security
54.62
Education
Status
47.58
Culture
57.63
Privacy
71.46
Safety
80.92
Social stability
72.30
Personel skills
56.79
Equality
84.25
Choices in life
Health
99.52
Primary social relationships
81.82
Secondary social relationships
43.49
Use of free time
50.56

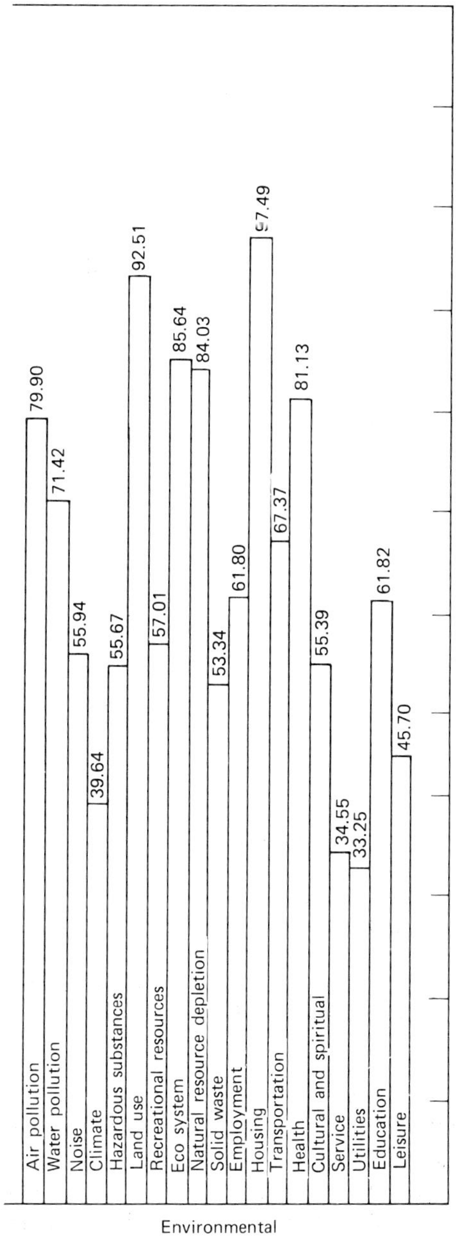

Air pollution 79.90
Water pollution 71.42
Noise 55.94
Climate 39.64
Hazardous substances 55.67
Land use 92.51
Recreational resources 57.01
Eco system 85.64
Natural resource depletion 84.03
Solid waste 53.34
Employment 61.80
Housing 97.49
Transportation 67.37
Health 81.13
Cultural and spiritual 55.39
Service 34.55
Utilities 33.25
Education 61.82
Leisure 45.70
Environmental

Figure 2

Political/Social, reflects the difficulties encountered in running the experiment and the disagreements that arose, and went unresolved.

That evening the weighting procedure was begun. The conference members were told first to allocate 100 points among the three major categories (Economic, Political/Social, Environmental) in order of importance. They were then told to allocate another 100 points for the factors under each category in order of importance. The results of this rating appear in Figure 1.

Because most participants split their votes at 5 point intervals, the distribution illustrates the accumulated score in every 5 point span (i.e., all scores between 32.5 and 37.5 are plotted at 35 to show the distribution). The total number of usable data points is 52 (28 of the 80 returns were unusable).

Many of the professionals hesitated to offer weightings; they seemed to feel that their opinions were worth so much more than the layman's that if they ventured any suggestion at all it would be taken as gospel.

The fact that the generation of a factor list and the weighting methodology was only an experiment did not seem to lessen the conferees' skepticism, nor did the fact that this was the first time any type of QOL conference had been held have any effect on their actions. The results obtained are undoubtedly biased; however, the degree of bias can never be ascertained.

Approximately 70 members turned in a QOL factor list with weights attached. The results were tabulated and the weighting for each of the factors gave us the ranking of the factors noted in Figure 2.

Analysis of Weightings

Important points to be made about the results in Figure 1 are as follows:

1. The distributions are remarkably smooth and reasonably close to normal.
2. The means are all very close to 33, which implies that there is no very strong difference among the components.
3. The confidence level for differences between the means is sufficiently small that the three components cannot be statistically distinguished as far as relative weight is concerned.

The ratings for individual factors on the list in Figure 1 were obtained by asking the participants to split 100 votes among the factors under each component. Because the number of factors under economics is smaller than the number under environment, for example, this scoring system tends to weight the economic factors more heavily than it should. To correct for this we multiply the direct votes by the *fractional bias* (the number of factors in the component list divided by the total number of factors in the whole list) for the component. The fractional biases (f), for the three components are:

$$f(\text{Economics}) = 0.26$$

$$f(\text{Political/Social}) = 0.34$$

$$f(\text{Environment}) = 0.40$$

The next step in the analysis was to weight each factor score with the weight assigned by the individual subject to the component under which the factor falls. Specifically, if we let C_i be the component weighting assigned by individual $_i$ to the component and F_i be the appropriately corrected (with f) vote assigned to the particular factor under component C, the weighted factor score for that individual would be C_iF_i and the average factor score for the group would be

$$\langle W \rangle = \frac{1}{N} \sum_{i}^{N} C_i F_i = \langle C \cdot \rangle F$$

The results of this computation (shown in Figure 2) illustrate some very interesting points. The most striking of these is the substantial spread in the weights of the sample population attached to the different factors. The 20 most highly weighted factors are listed in Table 3.

There are at least two significant observations to make about this list: (1) the factors on it are from among the objectively based social indicator types rather than from the class of psychological factors; (2) economic factors were well represented in the top ten list of factors, thus indicating that economic indicators cannot be ignored in developing a representative QOL index.

A decision analysis session was held on the last day; there, six QOL-type decisions were implemented:

1. A national policy establishes free public mass transit systems in the major metropolitan areas.
2. National water standards require complete recycling of all industrial and municipal water by 1982.
3. A national policy provides a guaranteed minimum annual income of $1500 per capita.
4. A national urban renewal policy is undertaken to provide for the replacement of all slum areas with planned communities consisting of mixed class housing by 1987.
5. A national energy policy requires use of low-sulfur fuels and atomic energy.

The participants were divided into groups of three each and told to rank the effect of the implemented decision on the QOL factors using a −10 to +10 scale (−10 being a complete and drastic negative effect of 100 percent on the QOL and +10 being a completely positive effect of 100 per cent on QOL).

Table 3. Top 20 Factors

Factors	Sum $(C \cdot S)F/N$
Democratic process	111.29
Public participation	101.73
Health	99.52
Choices in life	97.41
Housing	97.47
Economic security	96.95
Education	95.55
Land use	93.57
Essential living costs	92.97
Economic opportunity	90.65
Ecosystem	85.34
Equality	84.25
Natural resource depletion	84.03
Primary social relationship	81.82
Health (environment)	81.13
Safety	80.92
Income distribution	80.15
Air pollution	79.90
Income	74.43
Social stability	72.30

In the decision analysis session we wanted to learn whether any one or possibly a few of the QOL factors were drastically affected across the board with the implementation of every decision. If such a factor(s) existed then it (they) would be considered to be a predominantly influential factor to the decision and policy maker. However, meaningful data could not be extrapolated from this experiment.

Summary

The results presented in Figures 1 and 2, although interesting, certainly cannot be generalized in any way. The sample population used in this experiment was highly unrepresentative; hence any broad interpretation of the component and factor weightings obtained must be avoided.

This experimental conference was considered successful, if not for the generation of QOL factors and the weighting of these factors, then for the insights gained into the problem of QOL and into the decision maker's role and frame of reference.

The problem of developing a QOL matrix still remains; however, interest and impetus have been directed toward this topic area. The need is present, and only time and effort on by many workers will give us this most useful planning tool.

References

1. Raymond A. Bauer, ed. *Social Indicators*, MIT Press, Cambridge, Mass., 1966.
2. John Berenyi, ed., *The Quality of Life in Urban America—New York City: A Regional and National Comparative Analysis.*
3. Peter L. Berger and Brigette Berger, *Sociology: A Biographical Approach,* Basic Books, New York, 1900.
4. Peter L. Berger and Brigette Berger, "The Assault on Class," *World View,* June 1972.
5. The Brookings Institution, *Agenda for the Nation,* 1968.
6. Angus Campbell and Philip Converse, *The Human Meaning of Social Change,* Russell Sage Foundation, New York, 1972.
7. Peter A. Corning, *An Index for the Quality of Life,* Paper presented at the 138th meeting of the American Association for the Advancement of Science, December 1971.
8. Otis Dudley Duncan, *Toward Social Reporting: Next Steps,* Russell Sage Foundation, New York, 1969.
9. Joseph L. Fisher, "The Natural Environment," *Ann. Amer. Acad. Polit. Soc. Sci.,* May 1967.
10. Clare W. Graves, W. I. Huntley, and Douglas B. LaBier, "Personality Structure and Perceptual Readiness: An Investigation of Their Relationship to Hypothesized Levels of Human Existence," mimeographed paper, May 1965.
11. Bertram Gross, *Social Intelligence for America's Future,* Allyn & Bacon, Boston, 1969.
12. Martin V. Jones and Michael J. Flax, *The Quality of Life in Metropolitan Washington, D.C.—Some Statistical Benchmarks,* The Urban Institute, Washington, D.C., March 1970.
13. Robert E. Joyce, "Systematic Measurement of the Quality of Urban Life, Prerequisite to Management," *Seminar on Management of the City,* Research Analysis Corporation, May 1971, pp. 69–86.
14. Nake M. Kamrany and Alexander N. Christakis, *Social Indicators in Perspective,* International Seminar on Ekistics and the Future of Human Settlements, July 1969.
15. Abraham H. Maslow, ed., *Motivation and Personality,* Harper & Row, New York, 1970.

16. Arnold Mitchell, Thomas J. Logothetti, and Robert E. Kantor, *An Approach to Measuring Quality of Life*, Stanford Research Institute, Menlo Park, Calif., September 1971.

17. Research Analysis Corporation, *Environmental Quality Index—A Feasibility Study*, County of San Diego, Environmental Development Agency San Diego, Calif., June 1972.

18. Milton Rokeach and Seymour Parker, "Values as Social Indicators of Poverty and Race Relations in America," *Ann. Acad. Polit. Soc. Sci.*, March 1970.

19. Eleanor Shelton and Wilbert E. Moore, eds., *Indicators of Social Change*, Russell Sage Foundation, New York, 1968.

20. Nestor E. Terleckyj, "Measuring Possibilities of Social Change," *Looking Ahead*, August 1970.

21. U.S. Department of Health, Education, and Welfare. *Toward a Social Report*. Government Printing Office: Washington, D.C., January 1969.

Part IV

Role of Economic Analysis

Ralph C. d'Arge

Dr. d'Arge is Chairman of the Department of Economics, University of California, Riverside.

The central question posed by the sponsors of the symposium on regional environmental management was: Why has economic analysis not been more useful in regional environmental quality management? I believe this question should be revised to indicate, more specifically, whether economic analysis has been useful *where* it has been applied. The reason is simply that complete economic analyses have been applied only to a very few regional environmental problems, and in these instances they have generally been crash efforts for immediate decisions.* Thus we still do not know how useful economic analysis can be in the field of regional environmental quality management. But given the few *useful* environmental quality management studies now in existence, I would suggest that every scientific discipline's inputs are now in an experimental phase, including systems ecology and regional science. All are attempting to apply an existing methodology and perception to a fundamentally new problem—that of managing environmental quality within a comprehensive and consistent framework.

There is no paucity of literature in the field of economics dealing with environmental quality. More than 50 research articles and 10 books are published in the United States each year, and their primary focus is on regional or national environmental quality problems from an economist's perspective. Second, there is a significant contingent of economists in the evaluation branches of the EPA and other federal agencies directly concerned with environmental issues. Third, the methodological structure and method of application of economic principles to environmental quality problems is reasonably agreed on (with some significant exceptions) by professional economists. Thus most of the prerequisites for useful economic studies to be carried out for managerial decisions on environmental quality appear to exist. Then why has economic analysis not been a primary input and interpretive output for regional environmental quality management? I believe there are four major reasons: (1) the emphasis on regional economic modeling as opposed to regional economic analysis per se; (2) the advocate position and application by economists of economic value rather than "intrinsic value"; (3) the lack of successful integration and conceptual consistency between the economic analyses and other research studies focusing on a particular environmental quality problem; and (4) the expectation by decision makers and other researchers that economic analysis should provide more than a part of the whole analysis or should provide less than a part.

An adequate description of each of these reasons would require a research paper in itself. Here I shall only sketch their broad meanings. Not unlike other

* Some Exceptions include the studies of Hell's Canyon, the Alaskan pipeline, the Potomac River, and the Delaware River undertaken by Resources for the Future, automotive emissions controls under the auspices of Project Clean Air, University of California, and Rio Grande–Pecos Rivers at the University of New Mexico.

systems-oriented disciplines, when economists can't think, they model. That is, models are conceptualized that appear to be vaguely related to the environmental quality problem but do not directly or even partially solve it. The reason for such behavior is obvious. Models, particularly large-scale innovative ones, legitimize the research professionally, even though they may fail (and usually do) in providing sound policy advice. What appears to be needed is fewer resources devoted to traditional regional modeling (i.e., input–output tables) and more resources devoted to the economist attempting to conceptualize *and* distill an environmental quality problem down to manageable policy questions, options, and choices. Economic analysis is a learning, groping process and cannot be satisfactorily ritualized for all types of regional environmental problem.

Economists should recognize that in most environmental issues, nonmarket psychic and aesthetic values tend to predominate, thus emphasize their importance by way of concentrated research or at least recognition of them. The attempt to infer nonmarket values from observed market values should be an important component of the economic analysis.

Alternatively, others engaged in regional environmental quality studies should recognize that the social-environmental indicator with the broadest basis of public understanding is money. Thus some complex environmental tradeoffs may be better understood if the "measuring rod of money" is applied and sufficiently qualified.

Regional environmental quality plans or models should be developed by interdisciplinary teams that include economists from the beginning. Quite often, economists are only brought in after the pollution problem has been completely typified or modeled and it is too late to successfully alter plans to be economically compatible or consistent.

Policy makers expect choices to be delineated for improving regional environmental quality. Since economics is the "science of choice," often more is expected of it than it can produce. Economic analysis cannot provide *the* answer, but it can provide answers on the probable effects of various choices within both market and nonmarket contexts. Likewise, economic analysis cannot replace the inputs of other disciplines involved in regional environmental quality management research, but it can often place constraints or bounds on the scope and structure of needed research by other disciplines. The three chapters which follow, I believe, demonstrate how economic analysis can be useful in discovering choices and as aids in decision making. Donald McAllister illustrates a very innovative approach to understanding nonmarket measured demand for urban parks. Robert Anderson skillfully demonstrates some of the complications in pollution regulation when risk or uncertainty is an important ingredient of the regulatory problem. James Wilen rigorously analyzes the problem of valuing irreplaceable assets where such values might be expected to be unregistered in normal market processes.

The three chapters are representative of what economic analysis at its best can do for policy makers—identify problems and solutions consistently, identify (and sometimes measure) values that are important in decision making, and capture the important ingredients of a particular regional environmental quality issue.

Local Recreation Facility Location

DONALD M. McALLISTER

Summary

The traditional approach to planning a system of urban recreation (and other public service) facilities has been to apply standards that represent the judgment of experienced planners and administrators.[1] This chapter presents an alternative approach based on an economic analysis that makes explicit the objectives and constraints used to determine the "best" spatial pattern of local recreation facilities.[2]

More specifically, the chapter reports the results of research in developing a theory of and operational method for determining the "best" *size* and *spacing* of urban public service centers, using local recreation centers as an example. What size of individual recreation center and spacing between centers, feasible on a limited budget, "best" serves the residents of a region? At one extreme, all resources can be put into one, very large recreation center in the middle of a region; at the other, resources can be distributed among a large number of very small and widely dispersed recreation centers.

Following traditional welfare economic theory, the notion of "best" is divided into two components: efficiency and equity. The operational definition of equity chosen for this study is equality of opportunity as measured by variation (between people) in distance to the nearest recreation facility. Using this definition, the most equitable system would be one in which there is a recreation center next door to every home.

Dr. McAllister is an Assistant Professor in the School of Architecture and Urban Planning at the University of California, Los Angeles.

Efficiency is defined in terms of attendance at recreation centers, the most efficient system being that which maximizes total attendance in the system. There are opposing forces affecting attendance levels along the size-and-spacing spectrum. The major forces favoring a system of large, distantly spaced centers are scale economies and the greater attractiveness of large-center services. The system of small, closely spaced centers is favored by greater accessibility. The interaction of these forces yields a unique optimum size and spacing combination.

This economic analysis of location patterns is applied to a system of recreation centers in south-central Los Angeles. The tradeoffs between efficiency and equity as measured in the south-central community are discussed. The chapter concludes with an assessment of the usefulness of the methodology in recreation planning.

The methodology presented here is applicable to the optimal location of recreational and other facilities (e.g., clinics, schools) used by the general public. To the extent that the use of these facilities directly affects the quality of urban life, the methodology is relevant to the problems of environmental management.

The analysis primarily applies to services provided by public and quasi-public agencies which are distributed through a system of centers rather than a network. Hence the analysis deals with public facilities in one quadrant of a four-way classification scheme of public service facilities, as in Table 1. Although few public services in the United States fall entirely into a single category, most may be typified as predominantly one of the four. Facilities that are predominantly type I include libraries, hospitals, parks, recreation centers, schools, and day care centers. Freeways and public transit fall in category II. Type III services usually include police protection, fire protection, refuse collection, and postal services. Some type IV services are water, electricity, gas,

Table 1. Classification of Public Service Facilities

Method of Distribution Service	Characteristics of Public Service Facility: Service Centers	Network
Service not delivered to consumer (consumer must travel to service facility)	I	II
Service delivered to consumer	III	IV

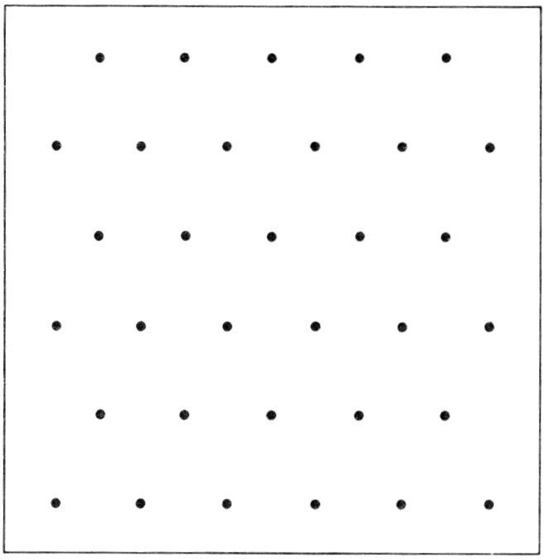

Figure 1. A system of small, closely spaced centers.

telephone, sewage, flood protection, streets, and sidewalks. The distinction between services that are picked up and those that are delivered determines whether the locational analysis should include a behavioral trip-making relationship, as in the former case, or a transport cost relationship, as in the later case.[3]

This study concentrates on one of the numerous dimensions to the problem of determining the appropriate size and location of local recreation centers (and other type I facilities); namely, the tradeoffs between a system of many small and evenly spaced centers and a system of one or a few large and evenly spaced centers, holding constant the budget for development and operation. Two extremes along the spectrum of alternatives are illustrated in Figures 1 and 2. Figure 1 is an aerial view of an urban region having a uniform popu-

Figure 2. One large center.

lation density and being served by a large number of small, closely spaced local recreation centers. Figure 2 shows the same urban region served by one large, centrally located center.

Our primary question here is, Within the bounds of a fixed budget, what size and spacing (i.e., distance between adjacent centers) of service centers is "best"? Following traditional welfare economic theory, the notion of "best" is divided into two components: *equity* and *efficiency*.

Equity

Equity can be defined in many ways. The operational definition chosen for this study is equality of opportunity as measured by variation between people in distance to the nearest recreation facility. Using this definition, the most equitable system would be one in which there is a recreation center next door to every home. Therefore, equity falls as the size and spacing of local recreation centers increase. In more specific terms, if variation in distance is measured by the standard deviation, it can be shown that equity is inversely proportional to the spacing or distance between adjacent recreation centers. Hence if spacing is halved, equity doubles.

Efficiency

Efficiency can also be defined in many ways. It is defined here in terms of the consumption of recreation services as measured by attendance at recreation centers, the "most efficient" system being that which maximizes total attendance in the system.[4] Other studies of public facility location have sought to minimize transportation costs rather than maximize the consumption of services. This approach is warranted when it can be shown that the minimum-transportation-cost solution also maximizes the consumption of services. However, the transportation cost criterion is inappropriate for the location of local recreation centers because it fails to consider trip-distance behavior and scale economies, which are discussed in the following paragraphs.

Opposing forces affect attendance levels among the alternative recreation systems lying along the size-and-spacing spectrum. The major forces favoring a system of large distantly spaced centers are scale economies and the greater attractiveness of large-center services. Economies of large scale are present in site planning (i.e., making the most use of a given unit of land), construction of recreation facilities, maintenance, and recreation supervision. The existence of these scale economies means that more recreation services can be offered (in the pure quantitative sense) on a given budget in a system of very large centers. The greater attractive power of large centers derives from the greater variety of recreation opportunities and the shorter average time spent by people in waiting lines. If scale economies and the attractive power of size were the only considerations in planning a local recreation system for an urban region, the most efficient system would contain one huge recreation center. However, numerous studies have demonstrated that use of a recreation facility is greatly affected by its accessibility, and a system of many small recreation centers generally is more accessible.

The effect of these factors is integrated in a theory of efficient size and spacing. The theory suggests that these forces generally interact in accordance with the graph of Figure 3. Thus attendance is relatively low in a system of very small, closely spaced centers, and it rises substantially with increases in size and spacing. In this portion of the curve, the greater economies and attractiveness of increased center size outweigh the effect of reduced accessibility. However, beyond some point, the balance is tipped in the opposite direction and further increases in size and spacing reduce attendance because of the overriding influence of accessibility. The most efficient size and spacing exists at the point at which service consumption reaches its peak.

There are two major steps in estimating the relationship in Figure 3. First we must identify the alternative size and spacing combinations permitted by a given annual budget for building and operating a system of local recreation centers. This requires information on the budget allocation to recreation and the relationship between center size and the (capital and operating) costs of providing recreation services. An example of alternative size and spacing combinations, taken from a study of south central Los Angeles, is given in Table 2.

The second step is to develop and calibrate a model that can be used to predict local recreation attendance as a function of the size and spacing of recreation centers. This estimation problem can be tackled with well-known

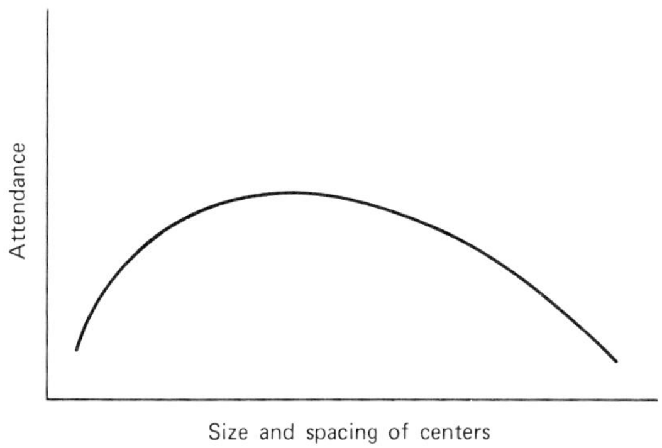

Figure 3. The most likely relationship between attendance and the size and spacing of recreation centers.

Table 2. Some Alternative Size and Spacing Combinations[a]

Spacing (distance between adjacent recreation centers, in miles)	Size (number of acres in an individual center)
1.0	0.01
1.2	1.68
1.4	3.65
1.6	5.92
1.8	8.50
2.0	11.37
2.2	14.56
2.4	18.04
2.6	21.83
2.8	25.92
3.0	30.32
4.0	56.84
5.0	90.94

[a] Source: McAllister, "Evaluating the Size and Spacing of Urban Public Service Centers," Ref. 2.

transportation demand forecasting techniques (with some modifications), using origin-destination information for trips to local recreation centers which can be easily gathered in a survey of local recreation users.

A model was developed and calibrated using recreation attendance data for south central Los Angeles. The estimated attendance curve (Figure 4) reaches its peak at a spacing of 1.8 miles between adjacent recreations centers, which corresponds to a center size of 8.5 acres (per Table 2).

Tradeoffs Between Equity and Efficiency

Now we have information on the equity and efficiency implications of alternative size and spacings for a given community. To summarize, equity is inversely proportional to spacing and efficiency is as described in Figure 4. As usual, equity and efficiency are at odds: equity is highest and efficiency is low in a system of very closely spaced centers.

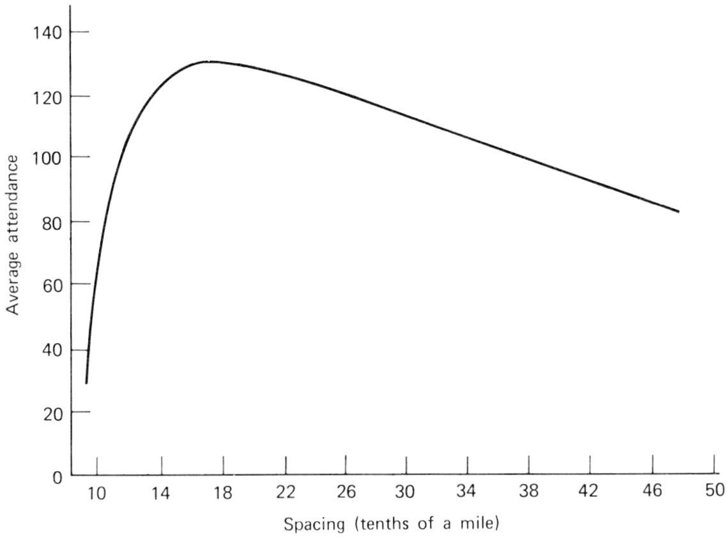

Figure 4. The relationship between average attendance and spacing, estimated from attendance data for south central Los Angeles. Source: McAllister, "Evaluating the Size and Spacing of Urban Public Service Centers," Ref. 2.

Unfortunately, there are no objective standards that can be used to assess the tradeoffs between the two criteria. However, an interesting characteristic of the attendance curve would appear to reduce the difficulty of this assessment: there is a fairly wide range of size and spacing alternatives within which attendane is not very sensitive. For example, the range of minus 5 percent from the optimal attendance level includes spacings of 1.5 to 2.4 miles and the respective sizes of 4.7 to 18.0 acres, which cover most current standards for neighborhood and community recreation centers. Similar results were obtained from a sensitivity analysis of the estimating equation, which means that the insensitivity is not unique to the case of south central Los Angeles.

Within the broad range of insensitivity to attendance, one might argue that the equity criterion should predominate. Clearly, a spacing of 1.5 miles is preferred to one of 2.4 miles, since attendance is the same in each and equity is 37 percent higher for the shorter distance. In choosing a spacing of 1.5 miles instead of the most efficient spacing (1.8 miles) one sacrifices 5 percent in efficiency to gain 16 percent in equity.

Ultimately, the decision of the "best" size and spacing is subjective. It will not be determined by an equation, but by people carefully weighing the important factors.

Conclusion

The analysis described in this chapter provides a framework for selecting and organizing information that bears directly on recreation planning decisions. It does not attempt to include all the factors that might be important in any specific situation; most factors vary from case to case. Since the analysis is designed to be generally applicable to a wide variety of situations, it focuses on two pervasive economic considerations. It is the responsibility of the decision makers to weigh them, along with other important factors in reaching a conclusion.

Notes and References

1. The analysis presented here is applicable to many public services but was specifically developed for planning a system of "local" recreation centers. The term "local" includes neighborhood and community recreation centers, which are distinguished from regional recreation centers primarily by their emphasis on playground activities.
2. The material is based on research that is more fully described in Donald M. McAllister, "Evaluating the Size and Spacing of Urban Public Service Centers: The Case of Local Recreation Facilities," Ph.D. dissertation, UCLA, 1970.
3. An exception to this rule is the case of primary and secondary schools, where the cost of transportation to and from schools is perhaps more relevant than trip-making behavior, since the receipt of services is generally mandatory.
4. This measure of efficiency represents a cost-effective rather than a benefit-cost criterion. The benefit-cost question of the optimal budget for a local recreation system is not considered because of the theoretical and operational difficulties in measuring the benefits of local recreation services. The Clawson method of measuring recreation benefits, which relies on information regarding the monetary cost of traveling to recreation facilities, applies primarily to regional recreation facilities to which the vast majority of users must travel long distances, involving substantial expenses for operating an automobile, bus and plane fares, and so on. The method is not readily applicable to local recreation centers because the majority who use these facilities either walk or ride bicycles to them, therefore incur almost no transportation expense.

Option Value and Irreplaceable Assets

JAMES E. WILEN

A particularly important offshoot of the recent rise in interest in environmental quality has been the increased concern with proper management of amenity assets such as unique natural environments and scenic attractions. On the national scale, recent controversies over the Alaskan pipeline and the Everglades Canal are cases in point in which opponents of development have argued persuasively that the uniqueness of the aesthetic experience provided by such areas warrants a very "special" type of analysis. On the regional level, similar pleas for cautious analysis have been advanced by individuals in support of preserving natural parks and areas of unique historical, geomorphic, or anthropological interest.

The decision maker charged with determining allocations relating to the preservation or development of such assets naturally wants to know exactly how such cases should be treated. What sorts of analysis ("special" or traditional) might be needed? One possibility is to use cost benefit analysis. An understanding of the conceptual basis is useful, however, before the results of such analysis are accepted with confidence for problems mentioned above.

Cost-Benefit Analysis/Willingness to Pay

The methodology of social cost-benefit analysis focuses on the social desirability (based, naturally, on some assessment of benefits and costs) of various allocations of resources that are held entrusted under the mandate of a

Mr. Wilen, a former Economics Fellow of Resources for the Future at the University of California, Riverside, is now at the Department of Economics, University of British Columbia.

public decision-making body. In a sense, cost-benefit analysis functions as a substitute, on the part of public decision makers, for signals that guide private decision makers to allocate resources under their commands. Just as individuals attempt, for given types of resources, to maximize their utilities in various alternative uses, public decision makers select various criteria (e.g., cost-benefit analysis) in an attempt to allocate public resources to maximize public or "social" utility (in theory at least).

The estimation of social benefits and costs resulting from a particular allocation is, of course, a task beset with difficulties. By making the initial assumption that social benefits are simply the sum of individual benefits, the task is simplified (at least in theory) to a micro level problem of examining how individuals value the outputs or services of a particular allocation of "public" resources. The task is also easier when outputs of allocations are marketed products. When outputs are brought and sold in regular markets, a ready measure of an individual's "valuation" of a good is provided by the price he pays for each unit, since he obviously is willing to pay at least that amount for whatever he purchases. Hence the concept of maximum willingness to pay is one measure of private valuation that has been adopted by economists. A more complete picture of individual willingness to pay is summarized in the individual's demand curve for a good (Figure 1). In Figure 1, the individual's demand curve for a particular good suggests that the measurement of the "value" to him of OX units is represented by the shaded area; that is, it is the maximum amount an individual would be willing to pay (unit-by-unit) for OX units. If the individual must pay a single market price of P_x per unit for the good, his maximum willingness to pay may be divided into his total expendi-

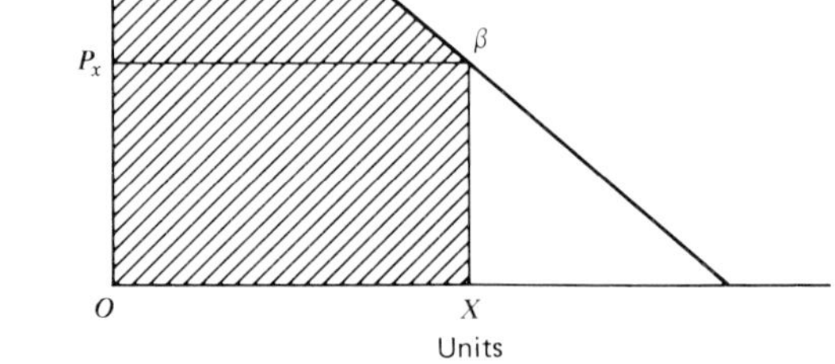

Figure 1

tures $OP_x\beta X$ and his consumers surplus $P_xA\beta$. Since a perfect price-discriminating monopolist is one who is able to extract the maximum amount for each unit, an alternative way of looking at valuation by individuals is to determine the amount a first-degree price discriminating monopolist could extract for (say) OX units. This would also be $OA\beta X$, since the price discriminating monopolist definitionally is able to capture the entire willingness to pay.

As we suggested in the opening paragraph, some recent test cases have questioned whether the "traditional" benefit estimation procedure based on the willingness-to-pay criterion is sufficient when the asset under consideration is irreplaceable. The problem is really twofold. In the first place, the demand for the services of such assets is uncertain. In general, individuals are rarely able to predict with perfect accuracy their future levels of demand for various goods. Although in normal circumstances this is not of much consequence to the individual, it does pose problems for the decision maker who must take an action *now* to assure future provision of a particular service. Accurate prediction and valuation are even more important when the commitment of the asset to one use is irreversible and it is known that no close substitutes for the alternative use will be available in the future. In the second place, even if demand were not uncertain, for many amenity services there exist no private markets, hence the estimation of values cannot be made using market demand curves.

The problem then becomes one of determining some "shadow prices" or proxies for market prices for services of these types. Various strategies have been devised for this purpose—ranging from questionnaires to rather elaborate studies of willingness to pay based on travel cost expenditures.[1] For the first problem, however, we have no theories of how benefits should be measured when demanders are uncertain, since the concept of willingness to pay has always been cast in terms of perfect certainty. In what follows an attempt is made to demonstrate how the development and application of a micro level framework might be used to answer such questions.

Willingness to Pay/Probabilistic Preferences

One could formulate a demand uncertainty problem of the type applicable here in several ways. A particularly appealing course is to assume, as in traditional microeconomics, that the individual makes choices as if he were attempting to maximize a utility function, except in this case we assume that he actually has a "split" utility function. Suppose that we divide time into a present period and a "future" period. Our focus will be on a situation characterized by the uncertainty of the individual's "future" preferences; that is, "something" beyond his control will determine whether he will demand the services of a particular good. This can be rigorously developed by assuming that the individual has two different future period utility functions corresponding to the events *demand* and *do not demand*, respectively. Only one of these events will occur in the future period, the result being that when the utility function corresponding to that event is maximized subject to the budget constraint, a demand of either positive or zero quantity of the service in question occurs. The maximization of the utility function corresponding to the event *demand* at various prices thus gives a demand curve of the type shown in Figure 1, which yields the maximum price that may be extracted from the consumer for every unit of the service in question, *if the event occurs*. The event *demand* will occur only with a certain probability of course; hence the demand curve as is cannot be used to compute willingness to pay for the uncertain demander.

What measure then might we use to determine willingness to pay when demand is probabilistic? The first candidate to suggest itself is some notion of "expected" willingness to pay, that is, the certainty of willingness to pay (given by the demand curve corresponding to the *demand* event) weighted by the probability that the individual will actually demand the good or service. Suppose that for our representative individual, the event *demand* will occur with probability π. Then under this expected value criteria, the expected willingness to pay to be derived from allocation of resources to the use in question would be:

$$E(WTP) = \pi WTP(P) = \pi[CS(P) + PX] \qquad (1)$$

where $WTP(P)$ = total certainty of willingness to pay (determined from *demand* event demand curve) as a function of the price charged

$CS(P)$ = consumer surplus derived from the *demand* event demand curve

PX = total revenues to be derived if *demand* event occurs

The rationale that supports using the measure of equation 1 is that it represents the expected value of the amount which a first-degree price discriminating monopolist could expect to extract from an uncertain demander. Unlike the certainty case, however, when uncertain demanders are present, the maximum amount that a monopolist may extract is dependent on the type of charge scheme employed. The foregoing scheme might be called an *ex post* charge scheme; that is, the total amount charged by the monopolist is collected *after* the uncertainty has unfolded. The estimate of that amount of course is made before the uncertainty has been resolved and is given by equation 1).[2] The price discriminating monopolist, however, has another possible pricing scheme at his disposal which may not be expected to yield the same revenues as the *ex post* scheme. The second pricing scheme what might be termed an *ex ante* charge scheme; that is, part or all of the revenues are extracted from potential demanders *before* they know whether they will be actual demanders.[3]

To examine the individual's response to such a scheme, consider the following: let utility in either the *demand* or *do not demand* situation depend on an amount X of the good or service in question and an amount M of another "all-purpose" numeraire good. Assume that the prices per unit of each good are P_x and 1, respectively. Let Y be income and RF be the reservation fee (explained below), both measured in terms of the numeraire good M. If the event *demand* occurs, the individual maximizes

$$U^d = U^d(X, M) \tag{2}$$

subject to $P_xX + M + RF = Y$. An equilibrium is reached in which

$$\frac{U_1{}^d}{U_2{}^d} = P_x \tag{3}$$

with $X > 0$ where $U_1{}^d = \delta U^d(X, Y - PX - RF)/\delta X$, etc. If the other event *do not demand* occurs, the individual maximizes

$$U^{nd} = U^{nd}(X, M) \tag{4}$$

subject to $P_xX + M + RF = Y$ and reaches an equilibrium in which

$$\frac{U_1{}^{nd}}{U_2{}^{nd}} \pm P_x \tag{4}$$

where $X = 0$. Note how the problem is set up. We are assuming for this pricing scheme that the individual must pay a *reservation fee* RF before the uncertainty is resolved for the *opportunity* to purchase the service, regardless of which event occurs.[4] Note that the reservation fee RF is paid even if the *do not demand* event occurs; hence the monopolist, to a certain extent, capitalizes on the individual's uncertainty regarding his demands by forcing him to pay something even though he may turn out not to be a demander.

The question of interest is now, What is the largest amount RF that may be extracted from the individual who is an uncertain demander? The answer depends on the stipulations of the reservation, the individual's wealth endowment, the anticipated price per unit of the service, and the individual's subjective probability of demand. We will assume:

1. Perfect exclusion is practiced; that is, those who do not pay the reservation fee are not allowed to purchase the good in the future period.
2. The individual's (certain) future endowment is $M = Y$ and $X = 0$.
3. The price per unit of X is fixed at P_x=0. This assumption will be relaxed later to show how RF varies with changes in P_x.
4. The individual's subjective belief that the event *demand* will occur is π where $0 \leq \pi \leq 1$.

Given the foregoing assumptions, the maximum RF can be seen to be the amount that will just make the individual indifferent between the two branches A and B of the decision tree in Figure 2. If the individual pays the reservation fee, his consumption bundle will be $(X, Y - RF)$ if he demands the good and $(O, Y - RF)$ if he does not. If the individual does not pay the reservation fee, his consumption bundle will be (O, Y) regardless of which event occurs. In terms of expected utility, RF thus solves it is the amount that makes the ex-

$$\pi U^d(X, Y - RF) + (1 - \pi)U^{nd}(0, Y - RF) = \pi U^d(0, Y) + (1 - \pi)U^{nd}(0, Y) \quad (5)$$

pected utility of paying the reservation fee equal to the expected utility of not paying it. The definitional relationship in equation 5 shows RF to depend on the probability of demand π, the amount of wealth endowment Y, and the amount of the commodity X. By assuming appropriate differentiability properties of U^3 and I^{nd} and by fixing π, Y, and X, RF is uniquely defined by equation 5.

The foregoing charge scheme, or at least approximations of it, are in actual use for provision of various types of services in the "real world." Many states have adopted reservation systems for state park use whereby individuals pay a reservation fee prior to the vacation season in anticipation of using a park space at a given time during the vacation period. Similarly, automobile clubs such as the American Automobile Association charge a membership fee that entitles individuals to free towing services, should the need arise. American Express and Diners' Club cards provide similar contingency services for a fee paid in advance of the use of such services. The difference between these cases and the one modeled previously is that the real world cases do not attempt to extract the *maximum* amount possible from each individual.

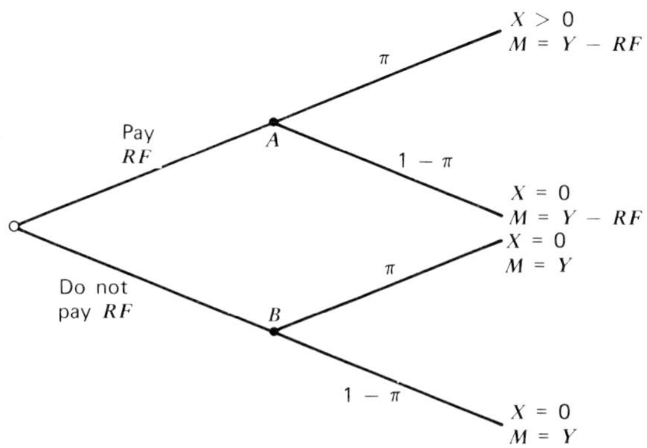

Figure 2

Other real-world charge schemes employ a combination of a reservation fee and a charge per unit of services demanded. TWA, for example, has recently instituted a "demand scheduled" flight. Under this pricing scheme, in anticipation of a cross-country flight several months hence, individuals may pay a reservation fee that entitles them to purchase their tickets at a substantial savings when the date of the trip arrives. If the individual does not demand flight services at the projected trip date, he forfeits his reservation fee. This arrangement may easily be incorporated into the earlier framework. If a positive price $P > 0$ is charged, we have as a modification of equation 5:

$$\pi U^d(X, Y - RF - PX) + (1 - \pi)U^{nd}(0, Y - RF) \qquad (6)$$
$$= \pi U^d(0, Y) + (1 - \pi)U^{nd}(0, Y)$$

Note the asymmetry in this formulation; if the event *demand* occurs, the individual's outlay will be RF for the reservation fee *and* $P \cdot X$ for the X units purchased and consumed, whereas if *do not demand* occurs,[5] the outlay will be only RF.

The significance of the *ex ante* models for cost-benefit analysis is as follows: when individuals have probabilistic preferences, the correct measure of expected benefits based on the amount a discriminating monopolist can extract is rather ambiguous. Equation [1] gives the expected gross revenues that a monopolist could extract under *ex post* scheme. Under an *ex ante* scheme of the type already discussed, the expected willingness to pay would be

$$E(WTP_R) = RF(P) + \pi PX \qquad (7)$$

Thus the expected gross benefits[6] under an *ex ante* scheme consist of $RF(P)$ (payable regardless of which event occurs) and the total revenues PX to be paid if the event *demand* occurs, weighted by the probability π that the event will occur. In general, the amounts given by equations 1 and 7 will diverge; hence we are presented with a dilemma, since we have two possible measures of the "value" of a service depending on what sort of charge scheme is used. It can be argued, however, the important variable is the maximum amount that can *actually* be extracted for individuals; hence the correct way to analyze the situation is to use the largest of equations 1 or 7, that is,

$$\max[E(WTP), E(WTP_R)] \qquad (8)$$

Comparing E(WTP$_R$) and E(WTP)

In view of the hypothesized discrepancy, the next issue to settle is the precise relationship between $E\ (WTP^R)$ and E (WTP). In particular, since the natural extension of "traditional" cost-benefit analysis would be simply to use the expected value of certainty willingness to pay criterion [equation 1], we need to know when $E(WTP_R)$ will be greater than $E(WTP)$. In other words, when will "traditional" techniques understate what could actually be extracted from individuals?

To answer this question, let us first consider the case of $P_x = 0$. We may state the following proposition.

> *Proposition 1:* If, for the representative individual, both utility functions are monotone increasing and strictly concave in the numeraire good,[7] and if in addition $U_2^d(X, Y - RF) - U_2^{nd}(O, Y - RF) \geq 0$, then $E(WTP_R) > E(WTP)$ for all $0 < \pi < 1$ when $P_x = 0$.[8]

In other words, if the individual's tastes and preferences can be described by utility functions with the foregoing restrictions, $E(WTP_R) = RF$ and $E(WTP)$ will look like Figure 3. If, for example, the individaul believes that the probability of demanding the good is π_0, he will be willing to pay a maximum of RF_0 reservation fee when $P_x = 0$. Note that when the individual is certain that he will not demand the good ($\pi = 0$), his RF is zero. Similarly, if he is perfectly certain that he will be a demander, he will be willing to pay some maximum amount λ for the reservation fee. λ of course is the individual's maximum certainty willingness to pay for the good, and since we are assuming $P_x = 0$, it is also the certainty consumer surplus derived from X. It is obvious therefore that the straight line joining $\lambda = WTP$ and the origin is equal to $E(WTP) = \pi\ STP$.

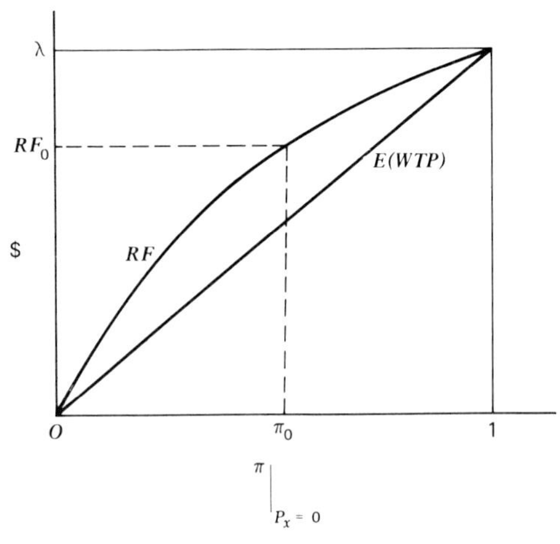

Figure 3

The foregoing may be extended to the case of P_x allowed to be positive as well. In that case we have the following proposition.

> *Proposition 2:* If for the representative individual both utility functions are monotone increasing and strictly concave in the numeraire good, and if in addition $U_2^d(X, Y - RF) - U_2^{nd}(O, Y - RF) \geq 0$, then $E(WTP_R) > E(WTP)$ for all $0 < \pi < 1$ when $P_x > 0$. Under these assumptions, $RF(P, \pi)$ can be shown to look like Figure 4.

For a given π, say π_0, the upper edge of the shaded slice traces out the (RF, P) combinations that are the maximum amounts extractable under an *ex ante* reservation fee pricing scheme. Thus for a given X, if the monopolist announces that the price per unit for the service will be P_0, there is a corresponding RF_0 which is the maximum amount that could be extracted as a reservation fee from the individual. Two remaining matters of theoretical interest are the choices of the optimal X and P that maximize net benefits. It would also be of interest to determine the differences in service provision that would be provided under each charge scheme. These questions go beyond the scope of this chapter, however.

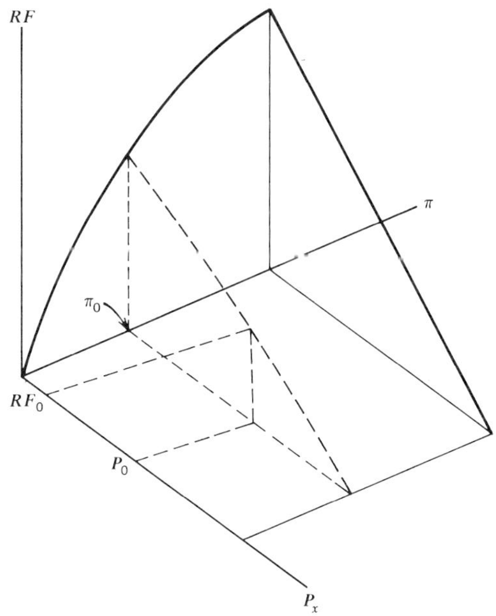

Figure 4

Option Value

The foregoing analysis lends some insight to the notion of option value, which has been debated for a decade. In 1964 Burton Weisbrod wrote an article in which he considered the following hypothetical situation: it is to be decided whether to close a particular natural park. Weisbrod's point was that even when total operation costs exceeded gross benefits (total willingness to pay), one could not necessarily decide in favor of closing the park because

> . . . there may be people who anticipate purchasing the commodity (visiting the park) at some time in the future . . . [and who] . . . will be willing to pay something for the option to consume the commodity in the future. This "option value" should influence the decision whether or not to close the park and turn it to an alternate use.[9]

Taken in context, Weisbrod was apparently suggesting that there may in some cases exist an additional type of benefit that might be important when deciding whether to irrevocably curtail the supply of a service.

Until very recently, the concept of "option value" was almost completely ignored by economists. Follow-up comments on Weisbrod's insightful (but nonrigorous) article were limited mainly to opposing interpretations of what Weisbrod "really meant" by the term "option value." More recently, attempts have been made to expand on the concept. Unfortunately, however, there still exists much disagreement regarding the meaning and significance of option value.

The notion of "option value" imparted in the quoted passage might be interpreted in several ways. For example, we might say that there are m individuals who are certain demanders and n more who are uncertain demanders. Weisbrod's suggestion then seems to be that the decision maker ordinarily takes into account the m certain demanders, even though he could also collect something from the n individuals for the option to consume. But this interpretation is surely not consistent with the way a rational decision maker would approach the problem, since a rational decision maker would also include some expectation of potential revenues that could be collected from the n uncertain demanders. The alternative interpretation is that Weisbrod was considering a charge scheme (applicable to all individuals) in which demanders are required to pay for the option to consume at some fixed positive price or at a zero price. It is this interpretation which has been most widely discussed in the literature that followed Weisbrod's original article.

The debate over option value has centered mainly on the question of whether the maximum amount an individual would be willing to pay for an option is less than, equal to, or greater than the expected value of consumer surplus. The reason for concern with this issue is fairly straightforward, and it relates directly to our discussion of *ex post* and *ex ante* charge schemes. Suppose that we are considering an individual whose demand for a service in the future will only occur with probability π. Let the certain future price per unit of the service be P and the quantity to be demanded (if the event *demand* occurs) be X. Then the expected value of the largest amount that could be extracted from this individual under an *ex post* scheme is, as we showed before, $\pi\ [CS(P) + PX]$. Now suppose that he were required to buy an option for an amount OV for the opportunity to purchase X units at price P. Under this alternative scheme, an *ex ante* charge scheme, the expected value for the amount that could be extracted from him is $OV + \pi PX$, *where* OV is the maximum amount he would be willing to pay for the option. The source of the debate about what option value really is thus becomes obvious. If, so Millard Long argued, "option value is exactly the expected consumer surplus,"[10] the two amounts $[\pi \cdot CS(P) + \pi \cdot PX]$ and $[OV + \pi \cdot PX]$ will be equal; hence the concept of option value (here taken to be the maximum amount a person will pay for an option) is rather redundant. Cotton Lindsay[11] was the first to counter Long's contention that $OV = \pi \cdot PX$. More recently, articles by Cicchetti and Freeman,[12] Byerlee,[13] and Schmalensee[14] have attempted to rigorously demonstrate the relationship between OV and $\pi \cdot CS$. Cicchetti and Freeman contend that OV exceeds $\pi \cdot CS$ for a risk-averse uncertain demander, whereas Byerlee reaches the opposite conclusion: namely, that a risk averter "tends to discount uncertain gains and pays less than expected consumer surplus." Schmalensee, using a state preference approach, shows that the relationship depends on the details of the individual preference relationships.

Our analysis is directly applicable to the option value problem because the reservation fee RF discussed is the same as the maximum amount an individual will pay for an option to consume. Hence we have one more proposition.

> *Proposition 3:* If for an uncertain demander with probabilistic preferences represented by U^d and U^{nd}, both utility functions are monotonically increasing and concave in the numeraire good, and if in addition, $U^d(X, Y - OV) - U^{nd}(O, Y - OV) \geq 0$, then $OV(P) > \pi \cdot CS(P)$ for all $0 < \pi < 1$ when $P \geq 0$.

Thus as a sufficient condition at least, whether OV is greater than, equal to, or less than $\pi \cdot CS$, depends on the specific details of the individual utility functions and the differences between them. For a decision maker facing a large number of potential demanders (say, n), it is not clear whether the inclusion of option value will make a difference unless he knows the details of each person's utility functions. For example, if the expected gross benefits are calculated using a nonoption *ex post* charge scheme approach, we have

$$E(GB) = \sum_{i=1}^{n} \pi_i[CS^i(P_i) + P_iX_i] \tag{9}$$

whereas, using the *ex ante* option scheme calculation we have

$$E(GB_0) = \sum_{i=1}^{n} OV_i(P_i) + \sum_{i=1}^{n} \pi_i P_i X_i \tag{10}$$

In general, unless all individuals are alike, $E(GB)$ may be either greater than, less than, or equal to $E(GB)$. Resolution of the problem thus requires empirical examination of the population of potential demanders for the services in question.

Summary

The attempt has been made to examine expected benefits when individuals are uncertain demanders. The concept of willingness to pay under certainty was reviewed, and the notion was applied to the case of individuals having "split" utility functions implying uncertain future demands. It was shown that based on the amount a discriminating monopolist can extract, the expected willingness to pay under uncertainty depends on the type of charge scheme employed. In the *ex post* scheme, all revenues are collected after the uncertainty unfolds; hence the monopolist is faced with the risky situation of having uncertain revenues. In the *ex ante* system, all or part of the revenues are collected before the uncertainty is eliminated. Three propositions were developed to suggest some conditions under which the amounts that a monopolist could expect to collect under each scheme were related. Finally, the notion of option value was discussed. The debate was reviewed, and it was shown that the *ex ante–ex post* charge scheme problem is formally equivalent to the option value problem. This being the case, the propositions developed in the previous sections apply as well to the option value problem.

References

1. See, for example, Marion Clawson and Jack L. Knetsch, *Economics of Outdoor Recreation*, The Johns Hopkins Press, Baltimore, 1966.
2. We are assuming that the decision maker in the present period is looking ahead into the future period and that he knows what the demand curve looks like but does not know whether it will in fact actually occur. He does know the probability of occurrence, however; hence he makes a decision based on the expected value or πWTP.
3. In this scheme, the individual still does not know which event will occur in the future period. He must, however, pay something to the monopolist in the present period (i.e., in advance of the unfolding of the uncertainty) to have the opportunity to buy if the event demand occurs.
4. It will be assumed that RF is "contracted" for in the present period but actually

is paid out of future period income. This assumption is made to ensure that total outlays for a good are coincident with consumption of that good in time.

5. Note that as in the case with $P_x = 0$, the reservation fee is paid in advance of the resolution of the uncertainty. The remaining amount $P \cdot X$ *is paid after the uncertainty has unfolded* (i.e., if the event *demand* occurs).
6. $E(WTP_R)$ represents the expected willingness to pay (gross benefits) under the reservation scheme.
7. This implies that $U_2^d > 0$, $U_2^{nd} > 0$ and $U_{22}^d < 0$, $U_{22}^{nd} < 0$.
8. The proofs of this proposition and those which follow are available on request from the author.

9. Burton Weisbrod, "Collective Consumption Services of Individual Consumption Goods," *Quart. J. Econ.*, **78,** (August 1964), 472.
10. Millard Long, "Collective Consumption Services of Individual Consumption Goods: Comment," *Quart. J. Econ.*, **81** (May 1967), 351–352.
11. Cotton Lindsay, "Option Demand and Consumer Surplus," *Quart. J. Econ.*, **83,** (May 1969), 344–346.
12. C. J. Cicchetti and A. M. Freeman, "Option Demand and Consumer Surplus: Further Comment," *Quart. J. Econ.*, **85** (August 1971), 528–539.
13. D. R. Byerlee, "Option Demand and Consumer Surplus: Comment," *Quart. J. Econ.*, **85** (August 1971), 523–527.
14. R. Schmalensee, "Option Demand and Consumer's Surplus Valuing Price Changes Under Uncertainty," *Amer. Econ. Rev.*, **62,** 5 (December 1972), 813–824.

Environmental Management Economics*

ROBERT J. ANDERSON, JR.

In spite of intensive scholarly study of alternatives to direct regulation, officialdom has continued undaunted to seek restoration and preservation of the quality of the environment by traditional regulatory means. Even as the number of pages of professional journals devoted to emission taxes, sales of emission rights, and similar subjects, has grown steadily, the paraphernalia of regulation—legislation, executive directive, and court opinion—have proliferated even more rapidly.

Withal, the work of the professional economist has had singularly little importance to the workaday practice of regional environmental quality management. This is most unfortunate, since economics has as much to say about the management of environmental quality by direct regulation as it does about the use of taxes or permits. Indeed, virtually every impediment to the success of the regulatory approach to environmental control can be traced more or less directly to economic causes. An understanding of the economic forces impinging on behavior under regulation is thus of the first importance in design and implementation of effective regulatory policy.

This chapter examines some simple economic principles that bear on the efficacy of such elements of environmental regulatory policy as emission standards and enforcement practices. The fundamental principle exposited is that in the absence of a widespread ethical precept forbidding acts that foul the en-

* This chapter is based on the author's more technical, "Some Economics of Quantity Regulation," presented at the Conference on Law, Economics, and the Environment, University of California, Riverside, February 1972.

Dr. Anderson is a professor in the Department of Economics at the University of California, Riverside.

vironment (similar to that which generally forbids murder or certain other crimes), regulations work only if effective penalties are imposed on violators. This is indeed a simple and an obvious principle, but many of its general ramifications are not sufficiently appreciated. Moreover, consideration of this principle in specific instances of the making of actual decisions about levels of standards, levels of enforcement effort, and so on, can provide some rough quantitative guidance.

The Exploitation Ethic

We shall operate under the broad assumption that there is no widespread environmental ethic, sufficient in and of itself, to ensure that the environment is protected in the absence of some kind of collective action. This seems to be a reasonable assumption, inasmuch as casual empiricism suggests that unfettered individual action in modern society is manifestly incompatible with a pleasing and healthful environment. In the absence of a special protective ethic, the environment is simply another resource for potential exploitation by consumers and businesses.

It is a principle—dare I say a law—of economics that this potential is realized whenever it is profitable for someone to do so. Individuals calculate the costs and benefits to themselves of exploitation, and they exploit whenever costs are exceeded by benefits. Because historically there have been no well-defined rights to many environmental resources, consumers and businesses have found it possible to exploit such resources freely. For example, we have not been made to pay for the air we use for combusting our gasoline because it is not clear to whom we owe such payment; accordingly, we have developed a transportation system that makes excessive use of this resource. In short, we exploit our environmental resources because it is individually profitable for us to do so.

Ecothug Regulation Model

Viewed in these terms, the fundamental problem of regional environmental management is to make it unprofitable to overexploit regional environmental resources. Whatever approach to management is adopted—be it the brainchild of an academic economist or a bureaucratic—it must satisfactorily resolve this problem if management is to be effective.

In developing some of the implications of this proposition for regulatory policy, it will be useful to formulate a simple economic model of the resource exploitation decision of a profit maximizing consumer or business. Because it is clumsy to write "a consumer or a business" each time we need some such phrase, let us lump together these two kinds of economic unit under the rubric "ecothug." We shall denote by "e" (for "emission," or "effluent," or "externality," or perhaps even "evil") the rate of exploitation chosen by the ecothug, and by $P(e)$ a function giving the profits to the ecothug associated with various levels of resource exploitation. It is assumed that this function is concave. The function $P(e)$ is depicted in Figure 1. The ecothug's profit maximizing level of resource exploitation is easily seen to be e.*

The objective of regional environmental quality management is to reduce the ecothug's rate of exploitation from e^* to a rate that is sufficiently low to protect the environment. To this end, the regional authority may prescribe a standard, providing an enforcement mechanism for detecting violations of prescribed standards and penalties for detected violations. Let E denote the standard defining the maximum legal level of exploitation prescribed for the ecothug, and let F be the penalty (monetary) per detected violation. It is assumed here for simplicity that the size of the penalty is independent of the seriousness of the violation.

In practice, detection of violations of standards is never perfect. There is always a probability, frequently a quite large probability, that a violation will go undetected. Let p_d be the probability that a violation is detected. It is assumed here, again for simplicity, that p_d is independent of the seriousness of the violation.

Under the regulatory regime just outlined, the calculating ecothug confronts profit possibilities distinctly different from those available when regulation is absent. In particular, he faces the decision of whether to reduce his rate of exploitation to a level consistent with the applicable standard, thereby certainly reducing his profits, or to continue exploitation in excess of the level to which

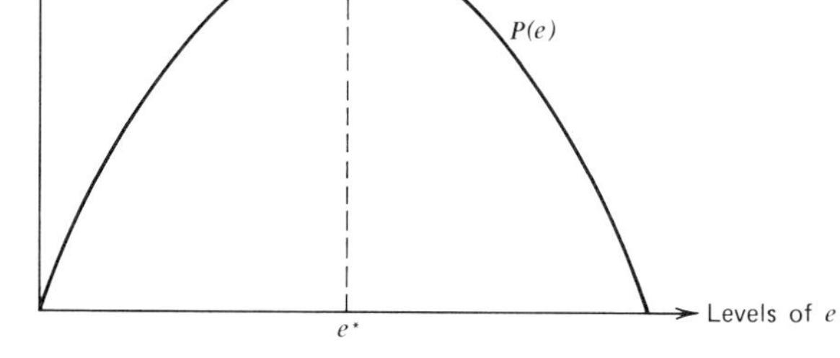

Figure 1

he is legally entitled, risking detection and fine. To predict which choice the ecothug will make, more needs to be known about his attitudes toward risks (i.e., how he feels about uncertain profits relative to certain profits). As a first approximation, one possible and probably not unreasonable assumption is that his choices are intended to maximize expected profits. Under the assumptions enumerated, the expected profits possibilities confronting the ecothug may be described by the two-part function:

$$P(e) - p_d^F \quad \text{if} \quad e > E \qquad \text{and} \qquad P(e) \quad \text{if} \quad e \leq E \tag{1}$$

Under the expected profits maximization hypothesis, the ecothug will elect that level of e which maximizes expected profits over the intervals $(0, E)$ and (E, ∞). The comparisons involved in making this choice are depicted in Figure 2. The maximum profits obtainable if the standard is to be obeyed are obviously $P(E)$. If, on the other hand, the ecothug were to be in violation of the standard, maximum expected profits are just as obviously $P(e^*) - p_d^F$. The choice between E (obeying the standard) and e^* (violating the standard) thus hinges on comparison of $P(E)$ with $P(e^*) - p_d^F$. If $P(E)$ is bigger than $P(e^*) - p_d^F$, the expected profits of obeying the standard exceed those of violating it, and the ecothug chooses to obey. If the reverse is true, the expected profits of violation exceed those of compliance, and the ecothug chooses to disregard the standard.

Simple though this model may be, it yields some interesting implications with respect to the regulatory decisions. It is evident, for example, that intensified monitoring (probabilities of detection) and/or increased fines for noncompliance render it more likely, other things being equal, that an ecothug

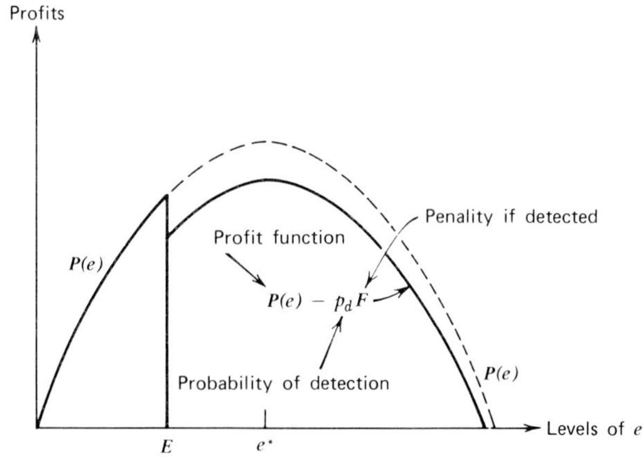

Figure 2

will conform to a standard. Only slightly less obvious is the effect of changes in the level of the standard adopted. Adoption of a more stringent standard (i.e., lowering of E) causes profits in compliance to decline relative to expected profits in violation. If stringency of the standard is increased sufficiently, an ecothug initially in compliance may be induced to violate the more stringent standard, with the counterproductive result that his level of resource exploitation may actually increase to e^*. In other words, making standards more stringent may actually make environmental quality worse!

The Real World and the Ecothug Model

The model presented is patently unrealistic in several respects. Most obviously, the assumption that detection probabilities and fines are unrelated to the seriousness of violation is unsatisfactory. Interestingly, however, the conclusions reached in the context of our simple model hold, albeit in slightly modified form, in many models complicated by these more realistic preconditions. It is thus generally true that regulatory decisions that increase the expected profits associated with violation (through poor monitoring, lowering of fines, or increasing stringency of standards) relative to those of compliance, tend to encourage violation and reversion to greater resource exploitation. In like manner, it is generally true that regulatory decisions that decrease the expected profits of violation (through intensification of monitoring, increasing of fines for violation, or relaxation of standards) relative to those of compliance, tend to encourage compliance.

These broad principles make it abundantly clear why environmental quality regulation has not worked very well to date. The monitoring effort has been grossly inadequate, and the penalties against the improbable unfortunate detected to be in violation, have been ridiculously small. In short, it has been profitable, usually by a wide margin, to simply ignore environmental regulations.

The principles stated also give guidance regarding what must be done if efforts to regulate environmental degradation are to be successful. In particular, it is clear that if stringent standards are needed to protect the environment, so then is vigorous monitoring and enforcement. Otherwise, standards become little more than lip service to environmental quality goals. Precisely how vigorous the monitoring and enforcement must be depends on the circumstance. If, for example, our simple model is accepted as approximately descriptive of the behavior of the ecothug and if the relationships between profits and standards are known, minimum expected penalties required to encourage compliance are easily computed. Some simple calculations of this type are presented in Table 1.

Although there is no hard evidence on current detection probabilities in

Table 1[a]

Ecothug	Control Required	Annual Reduction in Profits	Combinations of Detection Probabilities and Fines Sufficient to Offset Annual Profit Reduction			
			Fine and Probability		Fine and Probability	
500-MW power plant	Wet limestone injection	$1.3 million	$10,000	.36	$5000	.72
Plant boiler—360^6 Btu/hr	Switch to low-sulfur fuel	$1.0 million	$10,000	.27	$5000	.54
Lead smelter—96,000 lb ore/hr		$666,000	$10,000	.18	$5000	.36

[a] Costs and control methods are taken from the Implementation Planning Program, Office of Air and Water Programs, EPA.

enforcement of environmental regulations, it is probably reasonable to suppose that they are currently far less than any of the entries in Table 1, even those corresponding to fines of $10,000 per violation, which are quite large by today's standards. If this is true, much effort is needed to improve monitoring and enforcement to make it adequate to the task of deterring the ecothug.

Detection probabilities are largely determined by the monitoring technology adopted. Depending on the activity in question, monitoring may be done remotely or on site; it may be done continuously, at regular intervals, or on a spot-check basis; it may be done by highly automated equipment, or it may involve a large component of labor. In general, detection probabilities are highest with on-site, continuous, automated monitoring. Monitoring costs are also almost always greatest for this combination of options.

There are, of course, constraints other than cost which bear on the choice of monitoring technologies. The most effective technologies frequently are legally determined to constitute invasion of privacy or illegal search and seizure. As a result of cost and legal limitations, it would appear that for many types of environmental regulation, the prospects for materially increasing detection probabilities are not promising.

On the face of the matter, prospects for securing more effective regulation through increases in penalties associated with detected violations would appear to be substantially better. There are no particular direct costs associated with decreeing higher penalties and no apparent violations of constitutional strictures on investigation. Variability of this instrument is restricted only by notions of fairness and of what constitutes cruel and unusual punishment.

Some possible indirect empirical support for these general comments on possibilities for enhancing regulatory effectiveness through increasing detection probabilities and penalties is provided by the work of my colleague Morgan O. Reynolds on the economics of theft. Reynolds finds substantial statistical evidence that both these regulatory instruments are significant deterrents to crime of acquisition. Furthermore, Reynolds finds that it costs roughly between two and ten times as much per crime prevented to prevent crimes by increasing detection probabilities as it does to prevent them by sufficiently stiffening penalties.

Summary

It has been argued here that the problem of effective regional environmental regulation as it is currently being practiced is fundamentally a problem of devising a system of regulations and mechanisms for enforcement thereof which make it profitable, in purely economic terms, to behave in accordance with applicable standards. There is a strong presumption that regulatory mechanisms not so structured will not be wholly effective, or worse, will be outright failures.

A reinterpretation of some recent regulatory experience in federal programs for automotive emission control will perhaps serve to underscore this point. The federal program consists of certification that a limited number of prototype vehicle models meet federal requirements under rigidly controlled testing procedures. Production models, it is implicitly assumed, will meet the same standards because they are of the same design. Federal responsibility ends, for all intents and purposes, with the certification of a model for sale.

Unfortunately, most current automotive emission control technologies significantly reduce the performance and economy of operation of the vehicle. Many owners, who view these side effects as additional and unnecessary costs, have control devices rendered inoperative shortly after delivery. Inasmuch as there is no monitoring or enforcement after compliance certification of the prototypes, "disconnection" of emission control devices costs only the small amount needed for the simple servicing involved and results in rather substantial reduction in operating and maintenance costs, as well as "psychic" costs associated with a poorly performing automobile. Although we have no reliable numbers indicating the extent and trend of disconnections, there are rough indications that they are increasing as standards become more stringent and as the costs to the owner (of operating and maintaining emission devices) rise. This behavior is entirely consistent, it should be remarked, with our simple model. Note further that an implication of this model is that total automotive emissions might actually be reduced by adopting a less stringent standard if such standard reduced the perceived benefits of disconnection. This is indeed an interesting conclusion. The federal government and the states, recognizing the flaw in current practice, are moving not to relax standards, but rather to bolster detection probabilities by inspection of in-use vehicles, plus penalties for noncompliance. The model suggests that the latter alternative will be effective.

In sum, whatever the approach adopted to regional environmental quality management, economic analysis must be a key part of the management decision process. Management decisions that fail to mobilize economic incentives to work for, rather than against, the objectives of regional environmental quality management will simply not be effective.

Part V

Planning and Regional Environmental Management

Patricia A. Bonner

For years planning has meant regulation. Zoning, building inspection, highways, sanitation, flood control, water and public health, among others factors, were considered to relate to planning. Regulation alone, it has now been recognized, does not work to manage land and people. It is especially inefficient when each of the regulatory functions is separate from the others and/for further divided by jurisdictional lines or barriers.

Planning used to be for purposes of accommodating growth. Environmental management is leaning planning toward guidance of land use and development patterns. Today in many sections of the nation, planning is the tool used to channel and control growth. The difficulties faced when there is a policy to control growth are obvious and far reaching.

People have to live somewhere, and everyone wants his own area to be pleasant and uncrowded. Since the number of attractive places remaining near the population centers is limited, people are going further away from the metropolitan areas to live. But they still require and expect services. They want one-family homes, nearby schools and jobs, convenient shopping, good roads, and reliable water and garbage collection services. In short, they do not want to change their life styles when they change addresses. Higher density in the already developed areas would lessen the expense of providing the needed services, but then the American dream of personal home ownership would be destroyed.

Control growth, and the economic base changes—fewer new homes and industrial firms means fewer jobs and less in-migration. That cycle may bring temporary good, but what about long-term economic growth, the tax base, the jobs lost in materials, construction, and related contracting fields? What are the social costs of controlled growth?

We cannot accurately project the social, economic, and personal costs of controlled growth policies. We can, however, state that regional environmental management can work to integrate the many single-purpose entities; to examine how their functions relate and impact each other, the economy, and society; and to mitigate the detrimental economic and social impacts of environmental controls.

Planning and Environmental Protection

JOHN S. TOOKER

The Quiet Revolution

To explore the relationship between regional planning and environmental management processes, let us examine what is behind the great surge of concern about the environment. In their report prepared for the President's Council on Environmental Quality, Fred Bosselmann and David Callies declare that "a quiet revolution" is taking place across our entire country. They note that it is a legal revolution, whose leaders represent all political persuasions. The revolution involves the control of land use and the responsibilities for land use decisions.

In the mid-1960s, not only in California but throughout the country, citizens became aware of manifest changes in the environment. The air was foul, the water was polluted, and suddenly we realized that the vast supply of natural resources we had been utilizing for our own material gain, did in fact have limits. Our technology was increasing at such a rapid rate, our ability to understand the impacts of its by-products was not keeping pace. In our zest to stop air pollution, the means originally utilized to control hydrocarbons and carbon monoxide had the inadvertent result of a 50 percent increase in the emissions of nitrogen oxides from the internal combustion engine.

Californians began to realize that when there were only 10 or 15 million people in the state they had certain options and that the impact of their activities was much less far reaching than it was when 18 or 20 million people lived in the state. In California as the masses migrated for a number of reasons—economic, climatic, and social—they began spilling out of the urban

Mr. Tooker is the Director of the Office of Planning and Research in the Governor's Office, State of California.

centers into the prime agricultural lands of California so valuable for production of the fruits and vegetables, feed grains, and cattle. During this environmental awakening it became apparent that the environmental problems being experienced by California and other states were in fact people problems. Two alternative methods of coping with these problems were to control the number of people, and to control the uses the people made of the land to direct their activities to certain areas or to limit the uses of the land and preserve it for specified purposes or future options.

Regional Consciousness

At the same time, we began to realize that what man did in one area could have a great impact on man in another area within the same basin, airshed, or geophysical area. We began to recognize that environmental pollution and environmental problems did not respect political boundaries; therefore, methods of controlling pollution based strictly on local jurisdictions such as existing cities and counties proved to be inadequate in many situations. The need for regional bodies to oversee the activities in these geophysical areas or regions became apparent. As apparent as this need may have been, it raised a number of very serious questions, and many of them have not been solved to this day. The matter of home rule and our historic reliance on the municipalities and counties to represent us at the local levels is probably the most serious problem raised by the regional approach in planning. What degree of control should rest at the regional level and what should be left to the local level? Are the regional entities truly representative of the needs and desires of the local residents, or are these needs and desires best reflected by our existing cities and counties? This has historically been the case throughout our land.

Regional entities to control regional problems were created in the areas in which major crises developed. Examples of these in California are the state's participation in the Tahoe Regional Planning Agency, as well as the San Francisco Bay Conservation and Development Commission, and the referendum passed on November 7, 1972, which created a State Coastal Zone Commission and six regional coastal commissions. In all three examples regional entities were created to control land use, to save certain identifiable natural resources of the state. In addition, the use of land use control has been described by many as a preventive type of approach to avoid various types of environmental pollution that have appeared in other areas.

When regional entities such as those already discussed are created, more often than not they are single-purpose regional entities, not empowered or given the tools to do the types of comprehensive planning that would provide a preventive-medicine approach to environmental problems. In the case of the San Francisco Bay there is a proliferation of regional single-purpose entities;

all do their own planning, their own rule making, and in many cases their own permit issuing. What then is the relationship to the work done by the San Francisco Bay Air Pollution Control District and the San Francisco Bay Conservation and Development Commission? What is the relationship between these two entities and the various units of local government within their jurisdictions?

Special Districts

To further complicate the situation, besides various regional single-purpose entities and local jurisdictions, the special districts, which in the main are single-purpose entities, also cause some concern. For example, subregional sewer service districts are regulated by regional water quality control boards, but there is no overview on their total social, economic, and environmental impact. What is not looked at or considered in the planning process is the relationship of one special district and its functions to another. That is, we have special districts for providing water, sewers, schools, hospitals, and even cemeteries. When a decision is made by a board of directors of a water district to include an area under its jurisdiction, how does that decision impact on the functions of a sanitation or sewer district that provides sewer and water treatment facilities in the same area? At present, to the best of my knowledge, there is no structured coordination between the two; indeed, the decisions made by one will cause a reaction by another district, but not necessarily a coordinated, thought-out, planned action. I am not critizing special districts for the types of activity they undertake, but merely pointing out that the actions of these groups as well as the better-known organizations does have an effect on the total environment of any given region; therefore, their activity should be coordinated and planned in advance, along with the activities of all other organizations within that region, to ensure complementarity. Particularly in relationship to the growth-inducing or land use aspects of a project, the activities of all governmental agencies, whether state, local, or special districts, should be reviewed, evaluated, and coordinated. Thus a development that might have an impact on other areas within the region can be modified or eliminated to reduce the adverse environmental impacts, if any.

The South Coast Air Basin

As a specific example, let us consider the case of the South Coast Air Basin in the Los Angeles area. Without question, western Riverside and San Bernardino counties serve as a collection point for air pollutants generated in Orange and southeastern Los Angeles counties. The activities of various entities of both the government and private sector in these areas should be scrutinized and regulated to minimize the impact on Riverside. In the area, however, there are single-county pollution control districts, each regulating its own emission sources, except that the state preempts local jurisdiction in the case of mobile sources, such as the automobile. In 1970 in an attempt to bring together air pollution control units in a given air basin, AB 83, was introduced by Peter Schabarum, now chairman of the Board of Supervisors of Los Angeles County. The proposal called for the creation of a regional council within each air basin to be made up of a representative from the Board of Supervisors and the Executive Officer of the County Air Pollution Control District of each county in the basin. This was the first attempt made by the state to bring together all regulatory entities, as far as air pollution was concerned, to work on a coordinated regional basis to develop a plan to implement air quality plans for each air basin in California. Some progress is being made in the air pollution area, but it is becoming very clear that we must begin to look at the other environmental protection agencies and coordinate their activities. The State of California, for instance, has an Air Resources Board, a Water Resources Control Board, and a Solid Waste Board—all separate entities. Not unexpectedly, certain types of regulations adopted by one of these agencies can have very heavy impact on another board. For instance, suppose that the State Water Resources Control Board decides that an equitable means of eliminating a by-product from waste water treatment would be the burning of sludge. What expertise does the water board have to determine the effects of this action on air quality? At present there is no formal mechanism for an air resources board input into the decision-making process for a water board. Although there is some interaction among the staff of the boards, it would appear that greater review and coordination of proposed regulations is needed, ensuring that air quality considerations are taken into account during the deliberation of water quality regulations as well as solid waste regulations, and that water quality concerns are recognized and taken into consideration when the Air Resources Board and the Solid Waste Board consider adopting various regulations.

Regionalism and Local Government

But even if an effective regional agency for air, water, and solid waste were formed and began to adopt coordinated regulations, it would still encounter considerable difficulty in trying to implement specific plans. The authority to modify and plan land use currently is vested in local government, and local government can and does effect major changes in established land use patterns. Such changes executed in consideration of apparent short-term community interests can exert an unexpected and sometimes serious impact on the areawide environment. These effects may diminish the quality of the environment well beyond the boundaries of the community of jurisdiction. Yet in most cases affected communities on the periphery have no mechanism for representation in such decisions.

Unilateral action by individual communities may also impact on other areas within the region in an adverse manner, and the communities affected will lack the ability to make input into the decision leading to the adverse impact. For instance, in the San Francisco Bay region, several communities have recently passed what amounts to no-growth or stabilized growth policies, thereby limiting the amount of development that can occur in the community. How do these actions by a scattered few communities impact on the remaining communities which have not adopted such policies? Will certain communities in the Bay Area be impelled to accept increases in population, thereby creating a detrimental impact on their immediate environment while preserving the environment of those communities who were the first to pass stabilized growth or no-growth policies? Does the location of these communities in a certain portion of the region mean that additional growth by them will create air, water, or solid waste problems for the entire region? In the absence of an overall regional or umbrella agency concerned with land use and population as well as environmental concerns, these decisions will continue to be made on a piecemeal basis until another crisis mounts. Then we will be forced to rush in and deal with the problem when it is in an acute stage instead of trying to cope with the situation now through proper planning at the preventive-medicine stage.

Benefits of Regional Planning

Regional planning can have benefits far greater than mere environmental benefits. For example, in the efforts currently underway in the Office of Planning and Research, it has become obvious that California's foothill areas contain a large mass of land that is being subdivided into second-home or recreational, subdivisions. These units are considered to be used by part-time residents or in some cases by retired couples who seek to get away from the urban centers. My staff and I have begun to ask what the consequences will be if a four-day or even a three-day work week becomes standard, allowing a worker the option to live in the industrial complex during two or three evenings and his family to live in one of the foothill areas heretofore considered to be only temporary residences. Are the local jurisdictions that are allowing these developments to go on in their areas, providing adequate planning for police and fire protection, water, sewer, schools, park and recreation facilities, and all the other amenities commonly associated with the permanent home? The answer in the majority of cases is no.

Going beyond this, what will be the impact of something like the four-day work week and the migration to the foothill areas and the second-home subdivisions as just described, on the amount of tax revenues necessary within those areas to provide the services that will be demanded by the new residents. What will the secondary impact be if some of these subdivisions start to be converted to permanent residences? What kind of growth will accompany them to provide the other services that will be needed—barber shops, take-out food stores, restaurants, beauty parlors, and so on? I seriously question whether the local entities that are allowing these developments to occur, have given any thought to the possible environmental impact of new towns or communities scattered along the foothill regions of California, let alone the impacts on their demand for goods and services, which will have to be provided by greater tax revenue.

Now if I may let my agricultural background show for a moment—and there still is agriculture in many of these areas—if sudden spurts of uncontrolled growth occur here, what will be the impact on the livestock industry that historically has used these foothills for grazing? Would it suddenly become essential for California to allow all its beef to be produced in Texas, Oklahoma, Colorado, Nebraska, and Iowa, and shipped by truck or rail into California? And if that is what we opt for, what will be the impact on the thousands of men and women whose livelihood depends on the meat-packing industry in California? What types of employment can be generated that are nonpolluting on the environment, like the livestock industry, and where will they be located? It would appear that a sound comprehensive regional plan could take most of these considerations into account and plan for the growth and development and the various facilities that may be required. However, this is just not happening. Here again it seems that we will wait for the crisis to arrive before we react. The problem faced by local government is the need for a mechanism that can resolve areawide problems by local action and initiative rather than relying on the state government for all the answers. Rather than becoming involved in local and interlocal issues, it is the intent of the state to support the home rule tradition by encouraging the development of locally based areawide programs capable of solving areawide issues.

State Planning Policy

The Council on Intergovernmental Relations has adopted in principle recommendations for a state policy on substate districting and areawide planning organizations. The objective of this policy is to reduce the cost of government by providing for a more efficient and effective local, areawide, and state partnership in dealing with common problems, and to provide a basis for cooperation with federal organizations in the total decision-making process. It is the further objective of this policy to secure a more efficient and effective partnership among local agencies themselves for resolving areawide problems and issues. It is not an attempt to create regional government nor to add an additional layer of government. It is a positive effort for developing mechanisms for local solutions to areawide problems, and for supplying the framework for coordination of functional and comprehensive planning by local, state, and federal agencies. This policy will also serve as a guide to state and federal agencies for a coordinated delivery of services. In addition, this policy means to define the relationships of areawide planning to the local and state planning process and to develop the framework whereby local jurisdictions can participate in policy development at the areawide, state, and federal levels. The proposed policy reflects current and anticipated state and federal laws and administrative orders regarding areawide organizations.

Local General Plans

In keeping with the legislative mandates handed down by AB 2070 in the 1970 session of the California State Legislature, the Office of Planning and Research developed the state's first *Environmental Goals and Policy Report*, which designates a number of areas as having statewide significance or critical concern, each one falling within eight environmental categories. Local entities can use this information and include these areas in the various elements of their respective general plans, or they can identify the regions as areas of local or regional concern and take the necessary steps to adopt the ordinances that may be appropriate to preserve the critical resource identified. In some cases the areas are also of critical hazardous concern, and here local agencies can adopt or recommend the appropriate ordinances to protect the health and safety of the residents of their area from the various perils involved, such as fire, flood, or geologic hazard. But there is much more to environmental considerations and environmental concerns than the natural resource side.

A major consideration within these planning efforts should be the work performed today by local government in developing circulation and housing elements for their general plan. Much can be done here, particularly if regional concerns and factors are taken into account by local planners by working together even in an informal arrangement. For example, in the case of the Circulation Element, the location of highways and freeways can be critical. They can be placed and/or designed in a manner minimizing the impact on adjacent residents from noise. The rights-of-way can be designed to avoid such areas of critical environmental and/or archaeological concern as Indian burial grounds, which may be highly valuable or in which rare and endangered species may live. The placing of the highway routes, taking into account native wildlife, their habitats and habits can benefit human beings as well. The California Division of Highways recently contracted for a study identifying the natural resources found along the U.S. 50 Highway Corridor Crossing the Sierra Nevadas between Placerville on the west and South Lake Tahoe on the east. Because of the material developed by the Office of Planning and Research and the State Department of Fish and Game, identifying deer migratory winter routes, the planners doing the study indicated the location of migratory routes along U.S. 50 and related this to the number of deer killed and accidents involving motor vehicles. A deer or other animal in the path of a fast-moving vehicle presents an immediate danger to the health and safety of any occupant of the vehicle, let alone the beast. If highway routes can be placed so as not to block or not to cross directly the path of wild animals, accidents involving wildlife and automobiles can be reduced, thus saving other species as well as people.

Housing elements can also be utilized by planners within a region, working together to place high-density areas in such a manner that they can receive adequate service from existing or proposed water distribution or waste water treatment facilities. Location of high-density areas as well as industrial sites should also take into consideration temperature inversion, topography, wind velocity, and other factors that bear on the production and occurrences of that air pollution phenomenon commonly referred to as smog.

An element not even discussed yet is the water and sewer element of the general plan. The water quality amendment of 1972 to the Clean Water Act provides for various types of substate planning, basin planning, and areawide planning, as well as total state planning. To obtain construction funds under the state–federal program, a public entity must conform to the basinwide and areawide plans as certified by HUD and EPA. California's massive basinwide planning program was scheduled for completion on June 30, 1974. The plan identifies present and potential areas of critical water pollution concerns and develops a priority construction list for necessary waste water treatment plants, thus allowing California to continue to meet water quality standards in the year 2000. Although formulated under the guise of a water basin plan, it is basically a comprehensive land use plan for the total residential industrial and commercial needs and land use characteristics of the various areas. It is important that this basin planning effort be interpreted into areawide plans to ensure that water distribution and waste water treatment facilities can be coordinated and provided on an orderly basis when and where needed, and to facilitate the identification and correction of other potential environmental problems.

Summary

A regional approach can help resolve problems before they become acute, as they are in the Southcoast basin today. Had such considerations been examined and the appropriate ordinances adopted prior to development in southern Los Angeles and Orange counties, it is conceivable that the air quality in San Bernardino and Riverside could be vastly better than it is today. I would hope that planners, polititians, and concerned citizens have learned their lesson and that these planning considerations can be taken into account throughout the State of California in the years to come. We must work together, we must plan together, and we must learn to live together—or suffer the consequences together. There are no mysterious vapors that pollute. We are the polluters. We must realize this and strive to reach the only environmental goal California can have—that is, to create and maintain a productive harmony between man and his environment.

Is Urban Growth Controllable?

HON. PETE WILSON

I want to share a few thoughts with you about regional government, about growth and what can be done (and should be done) to control it, and about planning on a regional basis.

A friend recently sent me the following unattributed quotation that caught my fancy as a fairly concise outline of what we are up against in the world of planning.

> Urban planning requires a lot of ingenuity, a touch of utopianism, a healthy respect for economic feasibility, a bundle of technical expertise (and that's where you come in), considerable citizen involvement (and we don't lack that), concern for the public interest, appreciation of individual rights, and even some old-fashioned political cunning.

The author of that quotation might have added "new-fashioned political courage" to his list of requirements to make it complete. (The more cunning, the less courage is needed.)

The quotation is significant in that it reflects an understanding of the complex issue of growth, just as a recent issue of the San Diego American Institute of Planners newsletter put it another way:

> . . . ambition, energy and personal ethics and initiative are not enough to insure success in the planning function. The role exacts total involvement. The planning aspirant must develop the cunning of a kindergarten teacher, the aplomb of a topless waitress, the elan of a bus driver. Then he must be able to project the simple and unaffected innocence of an actor in a television commercial.

New View of Growth

These observers—and many others—are discovering a new mood in the American people toward urban development. Traditionally we have placed relatively few restraints on urbanization, regarding it as a mixed blessing.

Mr. Wilson became Mayor of the City of San Diego in 1972.

That is a tradition we can no longer afford. There is sufficient evidence—not just in San Diego County, but everywhere where great pressures for growth are threatening relatively untouched expanses of land—that people no longer accept the traditional process of virtually unrestrained, piecemeal urbanization as desirable.

There is some comfort in the realization by many public officials that they can no longer simply shrug, sigh, and accept the inevitability of urbanization that does not benefit their constituents. I would guess that there are few public officials who govern an area whose citizens feel in no way imperiled by too rapid growth.

Urban Planning in San Diego

For a moment, let us confine our discussion to San Diego. Perhaps no other single issue so consistently dominated the affairs of this city government in 1972 as the combined measures taken to properly plan and control urban growth in San Diego. It was a year when I somewhat proudly say that city government assumed its responsibility for determining the future shape of the city, rather than being a passive witness to its continuing growth.

In many respects, San Diego in 1972 was a laboratory for an entire nation that has become dramatically concerned about the adverse environmental and tax consequences of uncontrolled and premature development on urban fringes.

It is unnecessary, I am sure, to go into detail about the problems we faced—and at least partly resolved—in Mira Mesa on the northern fringe of our city. Perhaps less well known was the series of actions the council took to rezone more than 40,000 acres of raw, undeveloped land to lesser densities. In this case, the council inherited a situation in which thousands of acres of land on our urban fringes—and subject to growing pressures for development—had been prematurely zoned for residential and commercial use, years before the land was scheduled for development in our General Plan.

The effect of the so-called downzonings was to give to this City Council, and to future councils, the power of review (through zoning control) to determine whether the proposed development is consistent with the General Plan, whether the proposed development is desirable for the area in which it is planned, and whether the cost of providing municipal services to this new development will be offset by the concomitant increase to the tax base.

Consistent with these downzonings, the City Council adopted a policy spelling out our intention to establish growth priorities among these largely undeveloped areas by permitting phased growth in undeveloped areas only after a total cost/revenue analysis.

I am grateful for the participation of the Integrated Regional Environmental Management Project in the development of data bases and computer models which will help us accomplish this goal, as well as enable us to more meaningfully participate, along with our sister cities in the region, in comprehensive regional planning.

City governments still lack the express authority to control the timing and location of new development. It is not because we have not tried to obtain it, however, we have. In my last year as an assemblyman, and last year in the city's legislative program, I have sought from the state the clear, express authority in subdivision law to control the timing and location of new development. I regret to say that contractors and real estate interests have prevailed in the legislature, and urban sprawl remains a problem.

Suffice it to say that I am proud of our individual efforts to control growth: not to stop it, mind you. For I do not believe that a no-growth philosophy is either realistic or responsible. But I do want to take the reins from developers and say that growth decisions should and will be made, in City Hall.

Comprehensive Planning in San Diego

But the City of San Diego is not an isolated island. I hope that our good neighbors, the other 12 cities in the county, and county government itself will come to share the obligation that we feel to control growth.

The reason is obvious: certain problems, especially those of environmental quality, do not respect the arbitrary and artificial political boundaries drawn by men 30 or 50 or 100 years ago. Rather, they are problems that conform to watersheds and airsheds. They are problems that all environmental specialists deal with: transportation, sewage and solid waste disposal, air and water pollution control, and other areas of environmental concern.

Those participating in a regional environmental management conference know, and others are learning, that it makes sense to treat regional problems on a regional basis. It does *not* make sense to attack regional problems in a fragmented, uncoordinated fashion; and the jealous defense of small local sovereignties can offer no valid justification for attempting to do so.

Recognizing this fact, the City of San Diego is an enthusiastic supporter of the Comprehensive Planning Organization (CPO) of San Diego County, even though the county has chosen not to participate in it. And we are also looking ahead to the creation of possibly an even better vehicle, which would sufficiently serve the dual purpose of A-95 review of federal grant applications and proper cooperation and regional planning.

Last year the state legislative delegation for San Diego County held a day-long hearing on the need for regional planning and decision making and for intergovernmental cooperation. The hearing focused on CPO as a regional planning mechanism and the wisdom of converting from a joint-powers agreement (which is currently in effect) to a statutory binding regional decision-making process, which I personally favor. At that hearing, I also suggested that the time had come to look beyond the regional versus joint-powers question and to ask as well what should be the structure of government in the future of our single-county region.

Without prejudging a solution that should be recommended only after extensive investigation, I suggested that the legislature undertake such a study, applying to a model of San Diego County a number of governmental solutions that have been successfully employed elsewhere in the country. This would include application to our region of such consolidations as have occurred in Indianapolis, Nashville, Jacksonville, and Toronto. I have support for this idea from officials throughout this region. The legislature did, in fact, approve such a study, and I expect it will begin soon. It may take as long as five years to complete, but when completed, it should contain recommendations that can effect the economies and improve the quality and efficiency of service to taxpayers.

But more important, the recommendations of the legislature's study and their implementation could have very positive and far-reaching effects on our ability to plan and exercise land use decisions in truly regional fashion. I assume for a moment that the recommendations by an impartial and outside consultant, who had no ax to grind, would be for consolidation of governments within this region.

Assuming that this is the recommendation, and assuming even further that such a consolidation could be effected (and that is an *extremely* big assumption), the potential for easier and logical land use planning would be enormous. Ideally, the political bickering over land use matters would be considerably diminished, if not eliminated, leaving a far more hospitable climate for fighting the battle for our environment in the entire region.

Consolidation, even if that is the answer, is a long way off. Meanwhile, let me commend you for the job you are doing. For the new and exciting methods you are developing to make my job and the job of other policy makers that much easier in assessing the potential impact of man and his machines on his quality of life.

San Diegans are fortunate to live in a city, a county, and a region in which it is not too late to save the natural beauty we have inherited. Unlike so many other municipalities and counties that no longer have any place to grow, we are blessed with vast acres of undeveloped land— land that we can and must recognize as a limited and valuable natural resource deserving the closest scrutiny and protection before it is consumed. Because of this vast undeveloped land, we also face enormous pressures for development. As a matter of fact,

San Diego was one of the five fastest growing counties in terms of the number of new housing starts in 1972. And this situation challenges both the technician and the policy maker to work even more closely in the search for the proper balance of environmental and economic concerns that will guide our growth. I think we are up to the task.

So You Want to Limit Growth. . .

RICHARD RECHT

The case for controlling regional growth can be based on the argument that the increasing size of a region results in pollution of its air and water, congestion of its freeways, and other malign effects. Regions should therefore limit their population to a size that the environment can sustain without severe pollution and degradation. It can be observed that the size can be larger if more resources are devoted to technology that minimizes the pollution. Furthermore, no one can demonstrate the exact optimum size of the region that should exist, or the precise limit. These observations, however, do not change the basic argument that the environment cannot sustain the pollution resulting from unlimited growth and that it may therefore be in the best interest of the public to limit growth. This proposition has been enunciated by many people, and I presume it is generally accepted by environmental management practitioners that such limits are desirable and feasible.

Policies to Limit Growth

Steps to limit regional population growth are being taken in several situations. The regional plan for the Lake Tahoe Basin is one of the best known examples. A few months ago the Association of Bay Area Governments (ABAG) stated a given population as their policy. More recently political pressures resulted in their rescinding this limit and suggesting that it be considered merely one of the alternative plans the area could follow. On the state level, the 1970 California legislature required that the Office of Planning and Research produce a land use policy for the state. Implicit in this requirement is consideration of the population-holding capacity for the various regions.

Although limiting population may be desirable, its feasiblity is less obvious or, in other terms, the socioeconomic consequences of the policies necessary to achieve limited growth may themselves cause many problems.

Mr. Recht is the Senior Associate at Baxter, McDonald & Smart, Inc., consultants to the California State Office of Planning and Research.

What, then, are the possible policies for controlling regional population growth? We can call to mind the example of a gate at the fence surrounding the area. I believe this paradigm serves to encompass all policies under which the number of persons in the area is controlled directly by administrative means. This alternative is so contrary to our freedom of mobility that almost everyone would consider it anathema as a solution to our problem at this time. Another method of affecting population growth follows from the observation that regional location decisions are based on perceptions; thus modifying the perceptions of persons would result in changed migration patterns. The efforts of economic development commissions to lure industry to various areas is the most prevalent example of this policy. A somewhat perverse example is the informal campaign in Oregon to lie about the advantages of the state. Participating residents send their friends and relatives postcards painting a bad picture of the weather, job opportunities, and so on.

It appears to me that there is really only one way to stop regional growth in an attractive area. Stated very simply, it is to make the region less attractive to those who migrate to it. Thus if persons are migrating to an area because of its attractiveness, the only way to stop the trend is to make the area less attractive to them.

Job and Residence Quantity Controls

Having stated this logical observation, let us consider some of the policies usually suggested to control population growth. In general, they are based on the logic that population growth will be controlled if the number of residences and the number of jobs available are controlled. Among the problems involved in this approach, we can note that the policies affecting jobs and housing are usually under the control of municipalities. Thus in the Bay Area, although Marin County communities might have the tendency to say no, other communities with fewer conservationists among the electorate would be likely to welcome new industry and housing, particularly if it appeared that such growth would enhance the tax base. Clearly, controlling population through the control of residences and jobs must take place on a regional basis and will be possible only when there is adequate strength at the regional level of jurisdiction.

Let us consider the consequences of controlling the number of jobs in a region. Although difficult, it might be possible if we recognize that the only jobs that need be controlled are those that might be termed "economic base" or "export industry" jobs. If these are controlled, the service sector will take care of itself. The consequences of these policies is that the region will become unattractive because unemployment will be high. There will be job shortages, which will result in a decline in wage rates. To what extent will unemployment rise and wate rates decline? To the extent that the region becomes unattractive to those who would otherwise have migrated into the afea. It would appear that the burden will be borne by the employed of the region, probably lying heaviest on the poor and marginally employable. There would be a windfall to owners of existing industrial and office building sites at the expense of the owners of sites at which development becomes forbidden, although perhaps some of these profits could be taxed away.

An alternative policy is to control growth by causing a shortage of places to live. In this case a labor shortage results. Salaries rise until the area becomes less attractive to business. Given the lack of housing, the prices of homes and rents rise. The cost is heavy on the poor, but they are perhaps helped an equivalent amount by the job market. The retired and unemployed definitely suffer. There is a windfall to existing landlords at the expense of owners of now undevelopable housing sites.

It may seem that it would be possible to solve the situation by controlling both jobs and residences. I might first note that this probably implies much more control than we would welcome. The current supply of land for housing in the Bay Area is probably on the order of an additional 5 million population; the current demand is for land to accommodate approximately 50,000 persons per year. Consider the distortions in the market if demand remained a sizable portion of the 50,000 but housing was limited to upgrading and replacement by regulation and supply were only 20,000 units holding capacity per year—a 500-fold reduction. If we suffer all this control, we will still have in-migration until the combination of the housing shortage and the high unemployment rate makes the local situation so painful that people are no longer attracted to the area.

Government Land Controls

Some other observations can also be made about controlling regional growth. It is difficult to purchase all the land from which we would wish to remove the opportunity of development. As land is purchased, there would be a tendency for the remaining land to rise in price, reflecting the reduced supply. (Perhaps, however, this problem could be solved by purchasing land at its condemnation price as of the date of adoption of the regional plan mandating its purchase.) The alternative to the purchasing of development rights is zoning that disallows housing and job site land uses.

In this situation, the zoning usually becomes "zoning by existence." That is, by the time the problem has become so severe that people are willing to acknowledge the necessity of a solution, the desired size has already been reached, and essentially all undeveloped land must be zoned for open space, recreation, agriculture, or other nonhousing or job site uses. The Tahoe Basin plan is a good example of this phenomenon. The planners felt the desirability of keeping regional development as close as possible to the existing quantity.

Consequently, almost all unsubdivided land was zoned recreational. If two sites, only one currently improved, have a similar capacity for development, should they be zoned differently? Should someone who has paid taxes based on nonrecreational uses now be restricted to that use? Should someone who has paid on water district bonds now be unable to use the system? Obviously, equity problems are occasioned by such a policy.

I would note one good point in the midst of these gloomy observations. A study undertaken in the late 1960s in the Bay Area considered the fiscal consequences of the ABAG Open Space Plan in the Bay Area vis-à-vis the same population distributed in "urban sprawl." The study found that the savings in infrastructure costs—roads, sewers, water, gas lines, and so on—was of the same order of magnitude as the cost of purchasing the development rights to the majority of land in the Bay Area.

Finally, let us consider the possibility of a policy at the state level that would help to limit population growth in some of the basins of the state. Through facilitating the growth of alternative areas (new towns, enlarged old towns, etc.) difficulties must be admitted. New town programs have been well studied, and it has been shown that only a very small percentage of growth can be accommodated in such developments. A second question is whether they will draw people from the areas most threatened by their attractiveness or merely accelerate the exodus from relatively unattractive areas. A final problem: what means can be used to attract people to the area? Publicity has often been tried, and although it may have some effect, it is doubtful that it is very large. A more realistic program would be to provide a subsidy, probably in the form of public infrastructure. I assume the subsidy would be paid for by the state and devoted to less attractive areas with a larger population-holding capacity than they seem destined to achieve through natural market processes. We can see that if the subsidy were sufficiently large, some residents of more attractive areas might note the low cost of living in the area where development is being encouraged and emigrate to that area. But, as we can see, we have merely returned to the same old solution: make the attractive area no longer attractive vis-à-vis other locations.

A Personal Note

I hope my remarks have been provocative, but I do not wish to close without making my own position clear. I live in the Bay Area, and I wish to continue to live there. I find the area very attractive but fear that with 8 or 10 million neighbors we will have smog like an unnamed city south of us, we will have a dirty Bay, we will have to fight extreme traffic congestion, and so on. Of course, our population growth will eventually be limited regardless of whether we establish any policies to that effect. Remember, our basic postulate has been that when an area loses its attractiveness vis-à-vis other locations, in-migration will cease. In other words, if the pollution becomes sufficiently bad, the growth of an area will finally stop. I hope that we do not use this means to finally limit the growth of the Bay Area.

Physical Environment in Regional Planning

GARRY D. McKENZIE

Abstract

Historically man has consciously and unconsciously evaluated the physical environment in selecting areas of settlement. Geological controls have often determined urban sites; however, with the recent rapid increase in population, less desirable land is now being utilized. These developing areas are often selected on the basis of economic, political, scenic, and social factors, without heed of geological parameters. Resulting geologic hazards often necessitate services of the geologist—and they require large monetary expenditures to "manage" the physical environment.

This process must be reversed. Statewide or regional inventories of the physical environment must be undertaken as soon as possible and certainly prior to development of any area. Basic data maps (e.g., indicating slopes, surficial materials, mineral resources, and faults) must be prepared from existing geological reports and additional field work. After possible uses of the regions are determined, resource capability maps should be constructed using a "stop-light" code. Synthesis of these maps with information on biologic, cultural, and scenic resources, by a powerful planning commission, should result in rational determinations on the development or non-development of portions of the region.

A detailed inventory of the physical environment can provide planners with information necessary for making sound decisions on the development or non-development of a region. When the available physical resources and possible geologic hazards are known, wastefulness in resource utilization can be minimized. Standard geologic maps generally fail to communicate the physical

Mr. McKenzie is an Assistant Professor in the Department of Geology and Mineralogy at Ohio State University.

limitations of environments to planners who are unfamiliar with geologic terms. Such maps are now being used to supply basic data for the preparation of resource capability or suitability maps by geologists. Recent data from the first Earth Resources Technology Satellite (ERTS-1) suggest that extensive regional information can be obtained rapidly, thus enabling determinations of the physical and biological resources over large areas.

The importance of considering the physical environment in the habitation of a region can be seen through excavations of prehistoric man. The settlement of Prague indicates that man had respect for geologic hazards and preferred areas in which building sites were stable and water plentiful.[1] The eleventh-century inhabitants of this city found the high gravel terraces safe from floods and suitable for foundations and water supply. Even the "New Town" of Prague, built in the fifteenth century, utilized terraces that offered similar advantages. In the second half of the nineteenth century, however, Prague grew rapidly and without concern for natural conditions, and structures were placed in hazardous situations. Only in the 1930s did engineering geology maps become part of the urbanization process in Prague.

Geologic Hazards

Geologic hazards are naturally occurring or man-made geologic conditions that present a risk to life and property.[2] Although planning to avoid loss of mineral resources such as water, gravel, and limestone, is probably as important economically to a growing population as planning to avoid geologic hazards, the latter commands more public attention. The damage associated with geologic hazards, including landslides, earthquakes, tsunamis, subsidence, and floods, is visible to all; but the loss of a gravel resource covered by suburbia, or the pollution of a groundwater resource, generally goes unnoticed. Some geologic hazards can produce very significant resource losses; many could have been prevented through a combination of proper planning, sound engineering, and attention to disaster warnings.

During the 1970 earthquakes in Peru, debris avalanches killed 17,000 people. More recently the 1972 earthquake in Nicaragua destroyed Managua and was responsible for the loss of about 10,000 lives. In 1964 the high-intensity Good Friday earthquake in Alaska caused 114 deaths and more than $300 million damage. The Alaskan losses would have been much higher if the epicenter had not been in a relatively unpopulated region. In a populated region such as California, a similar earthquake could cause tens of thousands of deaths and extensive property damage.

Possibly the most destructive natural disaster in North America resulted from Hurricane Agnes in 1972. According to Parizek, 109 lives were lost, 360, 000 persons were evacuated, and damages could exceed $3 billion in Pennsylvania alone.[3] Floods, rockfalls, debris flows, and rising groundwater levels all caused damage. Much of this could have been prevented if zoning of flood plains had been enforced and building codes revised. People should be informed of the meaning of a 100-year flood; levels of former floods should be marked on utility poles and public buildings to encourage prospective buyers of property to check the possibility of future floods and the availability of flood insurance.

Preventive measures can also be taken to mitigate damage from earthquakes. Engineering geology maps that enable proper planning and engineering for urbanization are now available for a few cities in the world.[4] Seismic risk maps are also available.[5] If used in conjunction with suitable building codes in planning new structures, such maps could lead to the reduction of earthquake damage. Unfortunately, a suitable building code for public schools was not introduced in Los Angeles until 1933; as a result, older buildings are still in use that are unsafe. Schools that fall into this category should be replaced, but bond issues continue to be defeated. To complicate the situation, accelerations experienced during the 1971 San Fernando earthquake,[6] although subject to further interpretation, suggest that the present code may not be inclusive enough.

It has not been unusual for people to disregard the advice of trained observers. Two examples are the 1964 Alaskan earthquake and the floods in Florence, Italy. In 1959 a technical geologic report concerning the Anchorage lowland in Alaska documented the hazards of landslides that might accompany an earthquake because of the resultant instability of a clay formation.[7] This report never reached planners,[8] and it probably cannot be considered to have been a significant warning. The floods that have occurred in Florence are a different matter, however. Richard Klein relates the history of floods there, and the records date from 1117.[9] Since 1333 there has been a moderate flood in Florence every 24 years, a major flood approximately every 26 years, and a massive flood about every 100 years. Shortly after the 1333 flood, it was suggested that a reforestation program be initiated to reduce the floods that had developed in part because of deforestation. Nothing was done. In 1545 Michelangelo cautioned a nephew about buying a house in the Santa Croce district because the cellars became flooded every year. (Basements in this district housed priceless art treasures that were destroyed during the 1966 floods.) In the sixteenth century, designs were made for water impoundments in the hills, river dredging, and developments of chambers in the city to hold flood waters. These ideas were ignored. In 1797 a French engineer reported that the floods were a result of deforestation in the Alps, and 100 years later reforestation was implemented in several European countries. Floods decreased, but Italy ignored this result. In 1968 the debris in the River Arno from the 1966 Florentine flood had not been cleared, and the silt was 6 feet thick.

Resource Utilization

Geologic hazards are not the only problems that man has when he interacts with his physical environment. Another major problem is competition for resource utilization. Flawn[10] and Goldman[11] have suggested the advantages of multiple use as applied to the mineral industry. Basically this requires that the mineral industry be allowed to extract mineral resources, such as gravel, coal, limestone, and other material removed from the surface; later the land is converted to other uses, including housing, industrial parks, recreation, and even underground office space. Sports stadiums, warehouse areas, or theaters are suggested uses for depleted gravel pits. These uses complete the sequential land use. Simultaneous land use may also occur between the mineral industry and other users—for example, in the case of oil extraction and farming. Since mineral deposits of many types are geologic rarities, the concept of multiple use makes sense. Environmentalists have had good reason to complain for the lack of reclamation that followed extraction of minerals; generally this situation has improved—partly because it is profitable and partly because it is the law. Sometimes there may indeed be a conflict of interest that cannot result in multiple use. One case would be a rare environment or ecosystem; another would be a need to preserve wilderness areas because these are resources in their own right and would be destroyed if mining were to occur.

Inefficient or overextended use of natural resources, as is the case with some water supplies, could result in serious shortages. For several thousand years, availability of water has been a key factor in the habitation of an area,[12] and only recently have massive engineering plans been devised to make water available almost anywhere. Such plans for long-distance transport of water have not always been accepted as the ideal solution for all. Possibly man should consider water as a limiting factor, as has been suggested for Indianapolis,[13] or as a precursor of other limiting factors in our use of the land. When the need for water in an area exceeds the supply available under current technology, the inhabitants are approaching the carrying capacity of the land in the area. Maps of expected water-deficient regions for the United States have already been prepared on 20-year increments to the year 2020.[14]

The foregoing discussion illustrates the importance of determining geologic hazards and mineral resources before development of an area. Other factors that should be considered are economic, political, social, and biological; but these are outside the realm of the geologist, whose expertise lies in preparing an inventory of the physical environment for use in planning.

Evaluating the Physical Environment

The first step in evaluating the capabilities of the land is to assemble basic data on the geology and hydrology of the region. This may be done by searching the

Table 1. Some Examples of Basic Data Maps

Mineral resources	Groundwater
Flood hazards	Erodibility
Slope	Excavatability
Surficial water	Permeability
Landslides	Subsidence
Bedrock geology	Depth to bedrock
Bedrock structure	Depth of water table

literature, records, and maps; by consulting with governmental agencies; and by field investigations, or by some combination of these, depending on the area. The information may then be presented on maps known as *basic data* or *factor maps*. These may be no more than conventional surficial geology maps, depth-to-bedrock maps, or even topographic maps; however, some of the maps may be more specialized, depending on the area and the possible uses that might be contempleted for the land. Basic data map types are given in Table 1.

Once the basic data of a region have been assembled, the possible land uses, or more generally *resource uses*, are determined. Whether the area is to become a suburb, new town, or industrial complex, it must be determined where the waste may be disposed, and where the areas from which water and building materials will be obtained must be determined. Recreational areas must be delineated along with other factors in development of land. After much investigation it may even be decided that the region should not be used for human settlement but serve instead for farming, grazing, forestry, mining, recreation, or wilderness.

If a resource use is determined, two or more basic data maps are combined to determine possible suitable areas for that resource need. The maps produced for a resource are known as resource capability maps (Table 2).

Table 2. Some Resource Capability Maps

Sanitary landfill	Forestry
Agriculture	Land reclamation
Residential development	Subsurface mineral extraction
Parks and recreation	Liquid waste disposal
Transportation	Reservoirs
Industrial complexes	Groundwater supply
Surficial mineral extraction	Utility lines

Both state and federal geologic surveys are now preparing basic data maps, and some are also making resource capability maps. Techniques of compilation and examples of usefulness are given by McComas;[15] Moser et al.;[16] and Pessl, Langer, and Ryder.[17] Whereas the above-mentioned studies apply to county-wide regions or larger areas, also being developed are numerous engineering geology maps describing the physical environment of specific urban areas for construction purposes.[4, 18]

Many of the resource capability maps utilize the "stop light" code to distinguish degree of suitability.[19] A red area on the map is interpreted as being unsuitable for a specific use. In the case of a search for sanitary landfill sites, the red map areas might be underlain by permeable gravels that could permit contamination of an important aquifer; on a map produced to show location of sand and gravel resources, the red area might be underlain by shale—definitely not a gravel resource, no matter how desperately a planning group would like to zone the area for sand and gravel extraction. Yellow places on the map indicate caution, meaning that the resource may be marginal or that it can only be developed with much care and expense. Green indicates that the area is suitable for the resource for which the map was prepared. In some cases variations of these three colors—red, yellow, green—have been employed, providing greater flexibility in delineating resource suitability.[20]

In the preparation of a landfill capability map for Madison County, Alabama, the state geological survey[15] utilized the following basic data maps: soil infiltration index, depth to bedrock, potentiometric surface, physical obstructions, and preempted areas. The resultant color map is rather expensive to produce, and like many resource capability maps it may have a limited life because of technological advances. For example, improved technology might make it feasible to place landfills that have carefully monitored, and treated effluent, in areas that were not considered possible landfill sites at the time of preparation of the resource capability map.[21]

Other conditions that may lead to changes in the usefulness of specific resource capability maps are social, economic, and statutory.[22] Thus it is necessary for planners to return to the basic data maps with the new criteria for resource use and to compile a new resource capability map. This updating process and production of new color maps could be very expensive; however, the problem has been solved to some degree by producing "throwaway" computer-drawn resource capability maps. A good example of this is the planning study done on the Lake Tahoe region. Basic data were put on computer tape and when a resource capability map was needed (e.g., for timber harvest regions with minimum erosion), it was generated by computer for about $60.[23] A somewhat similar use of computers has been made in the Northwest.[24]

Probably many planners and geologists who are producing resource capability maps will ultimately use the electronic process just described. The colored or patterned capability maps will continue to be prepared in the standard way for certain resources and areas. Several states are now computerizing their land and resource inventory, and the advent of the ERTS program should facilitate the collection and use of this type of planning information.[25]

Ohio is one state that is now using the capability approach to land use planning.[26] Some projections for urban growth in Ohio have been made that apparently do not include consideration of the physical environment. One example is the report by the Commission on Population Growth and the American Future, which shows Athens County as completely urbanized about the end of the century, as people continue to leave rural regions. Geologic hazards such as landslides[27] and floods are not unknown in this area, and the available space for urbanization without expensive protective engineering is limited. The selection of Athens County as a growth center may have been based on past trends and a need for an urban region in this part of Appalachia to gain benefits from economies of scale. It may be that the region is no worse than any other in the vicinity for urban growth. Possibly the strip mines will become urbanized; however, before any plans for an urban region become final, a detailed study of the capability of the physical environment to support the projected growth should be undertaken.

The need to acquire data for land use planning on a regional scale before embarking on national plans for urbanization is real. The population of the United States will continue to grow for another 30 to 40 years, even though it has now reached the birthrate for a stable population. Thus plans should be made to accommodate this growth, somewhere in the nation, with minimum waste of human, material, and environmental resources. One step toward this end is the development of regional plans based on the capabilities of the physical environment.

Utilization of Plans

If there is a sound technique for implementing the regional plan that has included the information on the physical environment, satisfactory development or nondevelopment of the region should occur. Man should then be able to live in harmony with the land, at or below the population capacity of the area. Unfortunately many regional plans are ignored or are so modified by developers that the time of the planners may have been largely wasted. Indeed, some of the best planners find themselves and their organizations under pressure as their funds are eroded by local politicians who are speculating on land that has shown up "red" on the residential capability map.

There are a few good signs that real control of development is coming. Delaware and California have passed laws protecting their coastal regions. Some cities have improved zoning enforcement and have used government funds to purchase land for controlled use. Charitable trusts and conservation foundations have also moved to protect special areas by purchase, and it has been suggested that states do likewise. In Vermont, where large unspoiled resources were threatened with unplanned land development, a commission was formed to insure optimum use of the resources of the state and development without destruction.[28] This is being done with a comprehensive land use plan and commission control over "any act which has an undue adverse effect on the public health and safety, or the right of people to enjoy an unpolluted environment." More of this type of action on a statewide or regional basis is needed. The rate of preparation of statewide basic data maps will be increased by the use of satellite imagery; however, detailed, on-site investigations will still be needed.

Hopefully the rate of preparation of regional land use plans and the development of regulatory agencies to prevent wastefulness in use of resources will be such that resources will not be squandered. Unfortunately most people, including some planners making choices for the future, do not consider the long-range view. The reasons for this may be economic or demographic, or they may stem from a belief that what is being planned has only a limited useful life. Such myopic views could lead to serious resource shortages. People born in 1973 will be retiring in 2033 and could live until 2063, yet there are very few plans that extend our use of resources beyond the year 2000.

A statement by T. Hoopes, deploring our lack of long-term planning, is worth repeating:

> Our difficulty is that as a nation of short-term pragmatists accustomed to dealing with the future only when it has become the present, we find it hard to regard future trends as serious realities. We have not achieved the capacity to treat as real and urgent as demanding action today problems which appear in critical dimension only at some future date. Yet failure to achieve this new habit of mind is likely to prove fatal.[29]

References

1. Quido Záruba, "The Influence of Geology on the Development of the City of Prague," 23rd *Int. Geol. Cong.*, **12** (1968), 133–144.
2. Margaret Gary, Robert McAfee, Jr., and Carol L. Wolf, eds., *Glossary of Geology*, American Geological Institute, Washington, D.C., 857 pp., 1972.
3. Richard R. Parizek, "Agnes, Aftermath and the Future," *Geol. Soc. Amer.* (*Abstracts with Programs*), **4** (7) (1972) 621.
4. Robert F. Legget, "Engineering-Geological Maps for Urban Development," *Geol. Soc. Amer.* (*Abstracts with Programs*), **4** (7) (1972), 576.
5. William B. Clark and Carl J. Hauge, "When the Earth Quakes, You Can Reduce the Danger," *Calif. Geol.*, **24** (11), **203–214** (1971).
6. B. J. Morrill, "Evidence of Record Vertical Accelerations at Kagel Canyon during the Earthquake," in U.S. Geological Survey and National Oceanic and

Atmospheric Administration, 1971, pp. 177–181. "The San Fernando, California, Earthquake of February 9, 1971," U.S. Geological Survey Professional Paper 733, 254 pp.

7. R. D. Miller and E. Dobrovolny, "Surficial Geology of Anchorage and Vicinity, Alaska," U.S. Geological Survey Bulletin 1093, 128 pp., 1959.

8. Ernest Dobrovolny, and Henry H. Schmoll, "Geology as Applied to Urban Planning: An Example from the Greater Anchorage Area Borough, Alaska," *33rd Int. Geol. Cong.*, **12** (1968), 39–56.

9. Richard M. Klein, "The Florence Floods," in Alfred Meyer, ed, *Encountering the Environment,* Van Nostrand Reinhold, New York, 1971, pp. 139–148.

10. Peter T. Flawn, *Environmental Geology, Conservation, Land-Use Planning, and Resource Management,* Harper & Row, New York, 313 pp., 1970.

11. Harold B. Goldman, "Urbanization and the Mineral Industry," *Calif. Min. Inf. Serv.,* **12,** 12 (1959), 1–5.

12. Bernard Frank, "The Story of Water as the Story of Man," in *Water, the Yearbook of Agriculture,* U.S. Department of Agriculture, Government Printing Office, Washington, D.C., 1955, pp. 1–8.

13. Arthur Mirsky, "Geology and the Quality of Life in Indianapolis," *24th Int. Geol. Cong. Abstracts,* 1972, pp. 396–397.

14. Commission on Population Growth and the American Future, *Population and the American Future,* Report on the Commission on Population Growth and the American Future, Government Printing Office, Washington, D.C., 186 pp., 1972.

15. Murray R. McComas, "Geology Related to Land Use in the Hennepin Region," *Illinois State Geological Survey Circular,* 422, 24 pp., 1968.

16. P. H. Moser, K. E. Vanlier, J. D. Hardin, A. F. Sartwell, and W. P. Wilds, "Environmental Geology and Hydrology, Madison County, Alabama," Alabama State Geological Survey Atlas Series 1, 72 pp., 1971.

17. Fred Pessl, Jr., William H. Langer, and Robert B. Ryder, "Geologic and Hydrologic Maps for Land-Use Planning in the Connecticut Valley with Examples from the Folio of the Hartford North Quadrangle, Connecticut," U.S. Geological Survey Circular 674, 12 pp., 1972.

18. Robert F. Leggett, *Cities and Geology,* McGraw-Hill, New York, 460 pp., 1973.

19. J. R. Quay, "Use of Soil Surveys in Subdivision Design, in *Soil Surveys and Land Use Planning,* Soil Science Society of America and American Society of Agronomy, pp. 76–87, 1966.

20. Soil Conservation Service, "Soil Interpretations Guide for Urbanizing Areas, Grafton Township-McHenry Co., Ill.," (map): Soil Conservation Service and McHenry County Soil and Water Conservation District, 1968.

21. George M. Hughes, "Hydrologic Considerations in the Siting and Design of Landfills," Illinois State Geological Survey on Environmental Geology, Note 51, 22 pp., 1972.

22. Fred Pessl, Jr., "Earth-Resource Data for Land and Water Management in the

Connecticut Valley," *Geol. Soc. Amer.* (*Abstracts with Programs*), **4** (7) (1972), 626.

23. Robert H. Twiss, "Uses of New-Generation Geologic Information in Regional Planning," *Geol. Soc. Amer.* (*Abstracts with Programs*), **4** (7) (1972), 695.
24. Dan A. Horn, "Making Environmental Geology Useful to Bureaucrats, Builders, and Buyers—The Computer Datagraphics Approach," *Geol. Soc. Amer.* (*Abstracts with Programs*), **5** (1) (1973), 59.
25. J. R. Anderson, E. E. Hardy, and J. T. Roach, "A Land-Use Classification System for Use with Remote Sensor Data," U.S. Geolological Survey Circular 671, 16 pp., 1972.
26. T. A. Wakeman, personal communication.
27. S. P. Fisher, A. S. Fanaff, and L. W. Picking, 1968. "Landslides of southeastern Ohio," *Ohio J. Sci.,* **68** (2) 65–80.
28. Elizabeth H. Haskell, V. Price, W. Matthews, R. Cook, J. Davidson, and R. Booth, "Managing the Environment: Nine States Look for New Answers," Woodrow Wilson International Center for Scholars, Washington, D.C., 445 pp., 1971.
29. A. Spilhaus "Ecolibrium," *Science,* **175,** (1972), 711–715.

Attempting Regional Management

DAVID J. HEALY

The present methods of environmental planning and management are lacking both in authority and jurisdiction. The splintered, single-purpose, localized approach to managing a region's environment has led us into the present quagmire. A case in point—Los Angeles.

The dynamics of a city occur slowly. Undetected factors operating on the city today will affect it tomorrow. When decisions are made by hundreds of organizations, the resulting perturbations are extreme. Apply this to a region, and the results are staggering. We are at the stage where more uncoordinated perturbations will result in disaster. Some of the distortions are represented by filthy air, inefficient transportation systems, vacant redevelopment areas, unnecessary flood channels, untimely development, and incredible amounts of duplication of data acquisition.

The Los Angeles Example

In Los Angeles County alone there are 395 special districts. The quantity of different, unrelated decision makers in these districts is astounding. From mosquito abatement to waste water treatment plants, the special districts focus only on one problem with a solution that does not have to account for its secondary ramifications on the rest of a county, let alone a region.

Editors' note: This paper represents the individual views of the author, not the viewpoint of the Environmental Protection Agency; it should not be quoted in any other context.

Mr. Healy is an Urban Planner in the Air and Water Program Division, Environmental Protection Agency, Region IX.

Splintered decision making occurs in three definitively regional areas: air pollution, water resources and water pollution, and land use–transport. Single-county air pollution control districts only complicate a complex condition by not recognizing that air does not respect political barriers. Water resources and supply cannot be handled at the discretion of an individual water district to decide the quantity for the future population for its small area. Waste water treatment facilities have the advantage of being most flexible, for the sanitation district may not allow the planning commission any flexibility in selecting the site where development is to occur after the pipe is in the ground. Land use and transportation decisions are made at two extremes, the local and state levels. However, both land use and transportation have regional ramifications on certain scale, and catastrophic results have followed the independent making of decisions in these areas. The nonaccountability of one actor impacting on another actor can lead to disastrous conditions in a region such as Los Angeles.

The health, safety, and welfare of the general public has been disregarded for the sake of home rule and local control on matters that are of regional significance—which means that only regional treatment can resolve and make any progress toward improvement of conditions.

The private market place and separate public decision-making bodies are in conflict over the public health, safety, and welfare. Permitting our urban settlements to expand (and expand because the property owner wants to build at higher profit) occurs at the expense of rebuilding in renewal areas and of renovation and remodeling in the so-called gray areas. Of course it is much easier for outlying suburbs to grant building permit after building permit while rats and cockroaches roam the older sections of our central cities. As development expands, air pollution expands and intensifies.

Most of the air pollution problem in the Los Angeles Basin comes from automobiles. How can 122 cities, 5 counties, 850 special purpose districts, an association of governments, a pair of coastal commissions, and a multitude of state agencies be individually accountable for all the facets of the air pollution problem, which affects everyone?

They cannot. Many agencies have either directly or indirectly aided in producing the conditions that exist in the area, and they are for the most part powerless in resolving the problem. The subject of air pollution and the future is so complex that it boggles the mind to think that any significant long-term progress will be made with the governmental arrangements that now exist.

Water

For an illustration of how a decision from one special district affects the environment in the long term and how it ties the hands of the planning department, let us consider at the Los Angeles City Sanitation District (one of the largest special districts in the county). The district developed a Master Plan for Wastewater Treatment Facilities several years ago with population projections that are now far too large. Since the district is not directly answerable to the city or county regional planning commissions' population research groups, updating of the waste water plan is noncontinuous.

To accommodate growth, the district will size facilities larger for an excess capacity. This oversizing, however, interferes with rationally consistent approaches for good planning between the location, sizing and distribution of sewer services, and population location and density provisions of the planning agency.

Land Use

Whether this situation is significant in relating air pollution to population growth and location is doubtful. The county regional planners have not thought of their land use general plans with air pollution as a factor in the development of their adopted plans. And if they did, what difference would it make if Orange County and the City of Los Angeles did not?

Land use decisions are officially made by the local government, but the macro land use decisions have been made *de facto* by bodies and individuals other than the planning departments and locally elected officials of general purpose governments.

The California Division of Highways and California Department of Water Resources (DWR) have probably done the most in making the major land use decisions. The routing and location of freeways was made by the Division of Highways. The DWR determined what quantity of water would be made available to enable the area to attain its present population and supply enough for future populations. Freeways have made possible a rapid expansion for suburban developments.

Apparently bent on following the rule of thumb; "where there is a demand, fill it," the local governments saw fit to say yes to everything. Reminded of air pollution, they say "We have no control over it!" Congestion means more and/or larger freeways. These highway decisions were made generally without objections from local officials, because highways are good. Freeways are good if you have carefully planned how they are to be used and what is going to occur around them.

For the Los Angeles air basin to meet the 1975–1977 Clean Air Standards, drastic measures will be required. The measures themselves could and can be minimized by scrutization of the water, sewer, transportation, land use, and development plans. But we cannot expect cleaner air if these plans are developed by many different agencies on many different premises.

City traffic engineers are mainly concerned about catering to the automobile population and making driving easy and fast. The Southern California Rapid Transit District (SCRTD) is operating and planning for a mass transit system which has always played second fiddle to the automobiles of the Basin. The SCRTD is operating under tremendous obstacles which will never allow success. If public transit is not available as conveniently and as fast as that available by private car, and if there are no incentives to use public transit (or disincentives to automobile use), failure is assured.

Why do Long Beach, Santa Monica, and Culver City have their own bus systems? They can guarantee the services demanded within their own confines without struggling with the huge undersupplied and undersupported SCRTD system. But at the same time, not all the residents from Santa Monica live and work in Santa Monica. The RTD does run local buses from Santa Monica to downtown Los Angeles, but the slowness of the journey and passenger near-exhaustion on arrival after the 60 stops on the 16-mile, 90-minute ride on surface streets is certainly a strong motivation to use one's own car for traveling to work.

If 80 percent reduction in vehicle miles traveled (VMT) is to be achieved to meet clean air standards, what are our mass travel needs, and who has the ability to accomplish them? How is future growth of population to be accounted for? What development pattern is pollution reducing in shape, and energy and time efficient for public transportation routes? City centers served by transit corridors seem to be appropriate, but how can the multitude of planning agencies and governments see that such plans are implemented when further expansion of undeveloped lands is still supported at many of the jurisdictional levels.

I believe that a better environment can only be assured by a regulating and planning regional agency's orchestrating all the factionalized decision-making groups in the region.

Each individual city or county has an empire unto itself. The objectives of the local community can and should be preserved, but not at the expense of the health, safety, and welfare of any other locality or region. Determining the level at which local control should begin is a bigger question.

If one community attempts a policy of limited growth as a way to reduce pollution, what good does it do if the next community has a policy of expansion and more growth? This example does not describe the possible motivations to halt growth without agreement with the rest of the region. It allows for some elitist or perhaps discrimination factors behind a no-growth decision. For without agreement of all the affected people, one locality's policy is worthless in improving the overall region's environment and resolving regional problems.

So far, I have examined regional problems brought on by decisions of local agencies and governments. These bodies do not shoulder the blame alone. The State of California passed the enabling legislation authorizing all these different groups to make decisions that have great spillover effects.

The federal government funds so many programs on a splintered basis that it too shares the responsibility for the establishment of numerous single-purpose groups. The nationally funded programs have regional significance but are set up under separate agencies.

In the Los Angeles region, only the Southern California Association of Governments (SCAG) is taking a regional approach. This voluntary association is established and composed of local governments. Under OMB Circular A-95, it is authorized to act as the regional clearinghouse authority, commenting on the impact on the region of any project that involves federal dollars. It can take no other binding actions, however, and until recently SCAG has seldom exercised the authority it does possess, for to do a proper project review requires a significant amount of staff time. Operating without sufficient staff, the agency can adopt no more than a haphazard approach to enhancing and protecting the region's environment. Moreover, to take an adverse position on a project jeopardizes SCAG's existence. For example, if Los Angeles County were subject to negative review for a large capital grant and as a result did not receive it, the county, the largest entity of SCAG, would threaten to drop out.

SCAG is now trying to solicit comments from the other agencies which have the technical expertise to do the review. On a recent waste water treatment project in Orange County, SCAG asked for responses from the Orange County Air Pollution Control District (APCD) and Planning Department as well as from the Riverside APCD, since these communities would be downwind receptors of air contaminants from the potential serviced population. The water management agency had made a population forecast without consulting either the county planning department or SCAG. As a result of this review, the population ultimately served will be 230,000 people rather than the 365,000 originally proposed.

The other regional bodies involved in the Los Angeles Region are three State of California Regional Water Quality Control Boards. They do not have local government representatives and operate quite independently of the political jurisdictions in the area. They have been somewhat successful in cleaning up the waste water of the region but have not concerned themselves with the secondary impacts of their decisions.

On February 16, 1973, the State Water Resources Control Board (WRCB) adopted new regulations regarding funding and planning for waste water facilities. The new regulations have a sizing formula that greatly restricts the capacity of new facilities in air basins in which air quality is critical. Planning for facilities will be based on a no-growth population forecast. Any increase in capacity above the regulations will be paid for pro rata by the districts themselves.

The local governments within the region have little recourse in these matters. Although the California WRCB effort to minimize the effect of water programs on air quality can be lauded, all the other facets that should be part of an integrated solution of the environmental problems of the Los Angeles Region or any other region in the state remain unconnected.

Some Solutions

The preceding cases illustrate the problems necessitating resolution of a region's problems from individual splintered efforts. Now, I would like to explore solutions, the possible regional arrangements that might resolve the conflicts that exist in the present system's attempts at regional environmental management.

One Agency With Power to Act

All activities dealing with the regulation and planning of water, air, and land should originate from one regional organization. For example, water supply, quality, cleanup, and management would be the responsibility of one body. The regional organization for land would be responsible for the development of a land use and transportation plan aimed at reducing environmental conflicts, cleaning up, and maintaining environmental standards. All air pollution regulations, strategies, and permit authorities of the APCD would be the responsibility of the organization. These functions could be carried out in a flexible manner—the organization could act as an umbrella agency over existing districts, or it could take over the functions directly. Clearly the first approach would be a good first step in the development of a viable regional organization.

The regional organizations must be authorized by the state legislature to perform the specific duties listed. California's Council of Intergovernmental Relations proposed "Recommendations for a State Policy on Substate Districting and Areawide Planning Organizations" gives one view of the possible planning functions of a regional agency. It also outlines the structure and powers for such an organization. I do not believe that the proposal goes far enough in the implementation and powers aspect of the organization. The recommended financing scheme does not indicate how the funds will be raised, but there shall be contributions from local, state, and federal governments. Without explicit funding, no program will be worthwhile. Assured funds must be prescribed each year, whether by annual authorization, tax revenues, or some other method.

Local governments have attempted their versions of the regional approach (mostly at the threat of losing federal grant monies). It has been a weak approach, lacking effectiveness and support. Its mandated powers do not go far enough. For this reason, I believe that the state should prescribe the power and authority possessed by the regional agency.

The regional agency will become the superagency, exerting authority on air, water, and land. It would have the last word on permitting any alterations to the region's environment, including transportation patterns and modes.

Attainment and maintenance of all the environmental standards should be one of the regional agency's main objectives. It would have to be the information data base for all these areas, including demography. The agency would be responsible for making both long- and short-term decisions relating to the achievement of the environmental standards that exist. Planning would involve participation by the public as well as by local government.

The level of permit powers would be determined in accordance with land use–transportation–demographic plans. The agency would be responsible for the scale, density, shape, and pattern of development and redevelopment for the region, and for ruling on variances to such plans. Local governments would retain the local building permit authority and the authority to plan their communities in the confines of the regional plans.

Following are some of the other authorities the regional organization would have: the power to raise revenue through bond issues, and the regulatory jurisdiction of all operating agencies of special districts involved in facilities or programs that have planning and regulation activities conducted by the regional organization.

The regional organization would provide the guidance, permits, and direction for the special districts to carry out associated operating elements of regionally important programs. This structuring would assure that plans and capital improvements are consistent.

The regional organization would receive and review Environmental Impact Statements. The review function would allow the agency to appropriately put projects in regional perspective. At the same time the service as regional clearinghouse for federal funds would support this role.

The executive composition of the proposed regional agency could take several forms. An executive board could be composed of (1) locally elected officials of general purpose government; (2) a combination of locally elected officials and appointed public members; (3) a special board whose members are elected at large, preferably by regional district; or (4) a board whose members are all appointed.

In opting for a structure composed of locally elected officials or some combination including them, there would be assurance that agency decisions reflect the local government consensus. However, this consensus might not result in an improved environment, since plans and actions might be watered down before adoption. To help prevent that contingency, the rules and regulations of the organization should be so explicit that most issues can be decided merely by direct interpretation from the written rules and regulations.

An agency composed of all appointed members would have the advantage of fewer conflicts of interest in the decision making. The feasibility of such a structure would be dubious, however. California's Regional Water Quality Control Boards, for example, are set up this way. In granting land use and transportation authorities to a body consisting entirely of appointees, the state would be leaving the public almost helpless in the face of objectionable decisions.

The structure requiring separate regional district elections would assure that the representatives acted from a regional viewpoint. The regional organization would be their only governmental interest. Probably the fewest conflicts would be involved with this type of arrangement, and probably it would be the most efficient in terms of operations and responsibilities.

Only by orchestrating the powers of all those whose decisions can affect the regional environment, can we hope for environmental improvement. Actions to establish the regional organizations in urban and urbanizing regions are needed now. That this approach is vital is more apparent each day. We cannot continue the splintered approach to resolving conflicting demands on the regional habitat.

The Federal Role

In new and proposed federal legislation, we can see attempts to take a regional view toward the resolution of managing and planning the environment. The Federal Water Pollution Control Act has a section on areawide planning. The proposed Highway Trust Fund Amendments have two sections that are important to this subject: regional transportation planning would be done by an areawide organization responsible for planning, and EPA would have the yes/no authority on highways in cities. The proposed EPA Air Implementation Plan for Air Quality Control Regions calls for huge regional efforts to achieve and maintain the Clean Air Standards. The proposals for land use legislation, would be suitable for regional scale implementation, which could be a significant factor in achieving improved environments in the future.

A federal, state, regional, county, and local partnership would be necessary for the success of any regional organization. The guidance and shared experience of everyone's perspectives would further enhance the regional organization's effectiveness. The agencies involved would become more responsive to one anothers' needs and objectives. The many "us versus them" confrontations would potentially melt into the common bond of "us."

By insisting that localized interests see the implications of regionwide problems, EPA assumes a large role in fostering this new relationship. Most major environmental problems are regionwide, being related to poor land use, poor use of energy, inefficient transportation networks, and so on. In the regulation of the cleanup of our nation's air and water, tremendous steps have been and will be made to rectify these inefficiencies and wasteful practices. Most of our environmental problems have resulted from local decisions that should have been made at the regional level.

EPA's regionwide plans to clean the air by way of transportation, land use, and new source controls reflect the agency's attempt to solve a regional problem on a regional level. It should be noted that in the Metropolitan Washington, D.C., area these plans were generated by the Council of Governments there.

Under federal regulations, no waste water treatment facilities can be funded until an acceptable plan to treat and manage an areas waste water is developed. In California, joint federal, state, regional, and local meetings were arranged prior to planning on each project for fiscal 1973–1974.

The establishment of statewide land use planning to meet the requirements of the still pending National Land Use Policy and Planning Assistance Act would be one of the final pieces in the regional approach on air, water, and land use planning.

EPA has no requirements for an integrated regional planning body. However, the agency supports a consolidated planning approach, especially when it is accompanied by the mechanisms necessary to implement the plans that are made.

The Rural Town Before Suburbanization

JOE H. HALL

This chapter examines an attempt to build environmental awareness by discussing the financial costs of urban development in a small community, Tracy, California. The method utilized is unique not in concept but in its application to a "small town" that still has an opportunity to avoid the urban sprawl and pollution typical of other California communities.

A Small Town

One would imagine that the community of Tracy, California, belongs in the Midwest or in some quiet corner of the United States. The scene might seem to be idealistic to a harried big-city resident.

Tracy residents have little idea of what future years will bring. Perhaps it is unfair to say that the town is unaware of probable urban growth, but the magnitude and speed of probable suburbanization is hard to explain in everyday terms.

Why all the concern about the future? Located approximately 50 miles east of San Francisco, Tracy is the first town in the San Joaquin Valley adjacent to the Bay Area. The city is also surrounded by a ring of interstate freeways providing the new transportation links between the Los Angeles and San Francisco Bay Area regions. On this newly completed route it takes only 40

Editor's note: The opinions expressed in this chapter are those of the author and do not necessarily represent the opinions of the City of Tracy.

Mr. Hall, M.S., M.C.R.P., left the Tracy Planning Department, Community Development staff and is now with the Planning Department of the City of Santa Cruz.

minutes of relatively easy driving to reach the East Bay metropolitan area. Aside from the central location of Tracy, several other factors that have appeared in the last several years promise to greatly influence the city's growth. Environmental concerns have led many Bay Area communities to limit their annual growth rate, as is reflected in the Association of Bay Area Governments' recent policy statements on growth rates for the entire San Francisco Bay Area region. Thus it appears that limiting growth in the Bay Area will mean that new growth must come to Tracy, the nearest community with large land areas still available for development. Ironically, we are looking at the first case of urban sprawl caused by communities seeking to stop that very phenomenon. Of direct influence on Tracy, the communities of Livermore and Pleasanton, 20 miles to the west, have imposed building moratoriums until sewerage, water, and school facilities can meet the demands of the present population.

If planning considerations are creating the potential for suburbanization in Tracy, housing costs provide the direct inducement. It is still possible to purchase a new tract, three-bedroom home for $19,000 to $21,000, whereas a similar home in the Livermore and Pleasanton area would cost from $25,000 to $27,000. For a large number of middle-income people this substantial difference in housing prices might make the 20 miles of extra driving to Tracy seem much more bearable.

Aside from the factors that create the potential for growth, living in a small town can be a big change for the typical suburbanite. Life in Tracy often centers around long-established friendships or groups not seeking new members, and entertainment is limited; long drives are required to see a movie, attend a sporting event, or find other entertainment. Tracy's climate is not the most desirable, either—there are summer heats, winter fogs, and year-round winds. It is ironic that the factors that appear to be negative about a small town seem to attract increasing numbers of suburbanites who no longer feel any affinity to the culture or variety of urban areas.

The Economy

Agriculture is of prime importance to Tracy's economy. The city is surrounded by some of the richest farmland in California. The firms serving agriculture and processing its crops are the community's largest employers. Typical of this "agribusiness" are a tomato processing factory, a potato chip producer, and numerous farm sales outlets. Surprisingly, government is the city's second largest employer. A nearby defense depot, an atomic energy laboratory, and two aqueduct systems provide employment for approximately one-quarter of the town's working population. Another view of Tracy from the 1970 census reveals a picture slightly less optimistic than is indicated by its economic base. The town is losing its most productive age group, the middle-aged workers, while the over-75 and under-18 year-old age groups are increasing their percentage of the town's total 15,050 population. Other economic problems are associated with a high unemployment rate and the condition of a significant portion of the Chicano population who live below the poverty level.

Development Quandary

A review of Tracy's economy, attitudes, population, and politics reveals certain definite conflicts. Given a high unemployment rate and a tendency for middle-aged workers to leave because of apparent lack of opportunity, it is very difficult to oppose any type of growth that would assist the town's economy. Beyond this simple logic, it must be remembered that growth will decrease the amount of land available for agriculture, thus adversely affecting the town's prime economic asset.

Local politics also creates numerous problems. Most important, San Joaquin County has been traditionally lax in its review of urban development in the rural areas surrounding Tracy. Therefore, if Tracy chose to carefully regulate its growth, it is highly probable that a developer would simply avoid annexation and develop in the unincorporated areas near Tracy. This problem has perplexed Tracy for the last decade, and only recently has the county begun to change its attitude. Town politics also reflects a highly unpredictable attitude toward growth. Like many communities, Tracy is only interested in "clean industries" and "quality" tract housing. However, as might be predicted, actions have not always followed this philosophy. Although many in Tracy deride Oakland, San Jose, and other big cities, very little consistent attention has been devoted to ensuring that Tracy does not follow the same pattern.

With this general setting it is not difficult to imagine that the principles of compact growth, avoidance of urban sprawl, and advanced planning are difficult to sell. In Tracy a more direct method was needed: dollars and cents had to be directly associated with the problems of growth. A reason was needed to begin a thorough study of the development options available to Tracy, and this opportunity arose rather quickly.

Placing a Price on Urban Growth

The opportunity to thoroughly examine the growth options available to Tracy appeared when a 250-space mobile home park was proposed 3 miles south of the existing city area in the middle of a large walnut orchard. The size of this project surprised many town residents and convinced the City Council and Planning Commission that "something" had to be done. Since the project would have to utilize city sewer lines, the council declared a moratorium on all connections to city sewer lines outside city limits. Following this decision, consensus on the next course of action vanished. Several councilmen and planning commissioners felt that the Community Plan should be revised to reflect a smaller city area, whereas others believed that a simple denial of permission to connect to the city sewer system was sufficient. Recognizing this split and at the same time realizing that future growth was inevitable, the Community Development Department proposed that a study be conducted to determine the exact cost of future development, thus creating what might be called a "growth budget." It was agreed that this would be a "wise course of action" and would avoid the ambiguities of a new Community Plan. As is evident, the approach deals with environmental questions through indirect comparison. Instead of specifying the disadvantages of uncontrolled urban growth, the costs of growth study spells out its high cost, thereby building a case for carefully controlled growth which is less costly in both financial and environmental terms.

To focus the study on the financial aspects of new growth, the Community Development Department carefully selected three objectives that would provide both an idea of development costs and a program for controlled urban growth. The objectives were:

Cost revenue analysis of the south Tracy area, to furnish guidance to the city relative to policies regarding capital improvements, development fees, timing of services, and level and quality of development.

Evaluation of that portion of the adopted Community Plan as it relates to the area in relation to residential, commercial, and industrial land use.

Investigation and recommendation of necessary zoning changes to implement phasing of the Community Plan.

At its outset the Urban Growth Policy Discussion created a stir in Tracy. Previously, it had been assumed that growth would be in a north and northwest directions. The "south side" had long been considered to be an undesirable area, and citing the examples of Stockton and Fresno, many assumed that Tracy would follow suit and grow to the north. However, the Urban Growth Policy Discussion pointed out that the new interstate highway bypassing Tracy had been built as far north as possible. This distance meant that the northern portion of the city could accommodate only approximately one-fifth of the anticipated urban growth. It was further pointed out that growth to the northwest was limited because of a series of problems:

Lack of sewer facilities

Lack of drainage, and a high water table, which makes construction of a drainage system extremely expensive.

The existing property ownership pattern, which appears to preclude an orderly progression of growth.

This left Tracy with one alternative: defy tradition and grow to the south. Here the city had a sewer line with sufficient capacity to serve new residential growth without substantial investment or improvements. Since the south Tracy area also contains some highly productive agricultural lands, any discussion of growth in this area had to be formulated to avoid recommendations that might ultimately produce uncontrolled growth. To deal with this problem, the study immediately described how provision of municipal services to areas not adjacent to the city was measurably more expensive than servicing an area adjacent to the city.

To provide a more complete analysis of costs and alternatives available to the city, the study detailed the costs of municipal services and outlined areas in which growth might occur. In determining costs, six municipal services were chosen as being most closely associated with new development, these were: arterial streets, major sewer lines, major water lines, parks, fire, and police. Next, eight areas were selected as being likely to develop either because of present construction or the potential for development over the next 10 years (see Figure 1). These areas were then assigned a population based on their zoning as projected in the Community Plan. Completing the analysis, revenue generated by development in the eight areas was estimated, assuming that it would be approximately two years before revenue generated from each area was fully realized.

The last step in the analysis, a cost-revenue comparison, revealed that anticipated growth in each of the eight areas (plan phases) would increase Tracy's population to 25,000 and result in a deficit of $390,830.

Plan Phases	Expenditures	Revenue	Net Charge	Balance Forward
I	$ 48,683		−$ 48,683	−$ 48,683
II	96,410		− 96,410	− 145,093
III	494,150	$358,460	− 135,690	− 280,783
IV	518,765	389,668	− 221,097	− 501,880
V	351,853	399,338	+ 47,485	− 454,395
VI	837,525	507,303	− 330,222	− 784,617
VII	335,492	573,309	+ 237,817	− 546,800
VIII	446,451	602,421	+ 155,970	− 390,830
	$3,121,329	$2,730,499	−$390,830	−$390,830

When considering the foregoing deficit, it should be remembered that overhead costs involving planning, finance, and engineering were not included. In addition, school costs were excluded because at that time it was felt that court cases affecting school finance might substantially alter current estimates. The

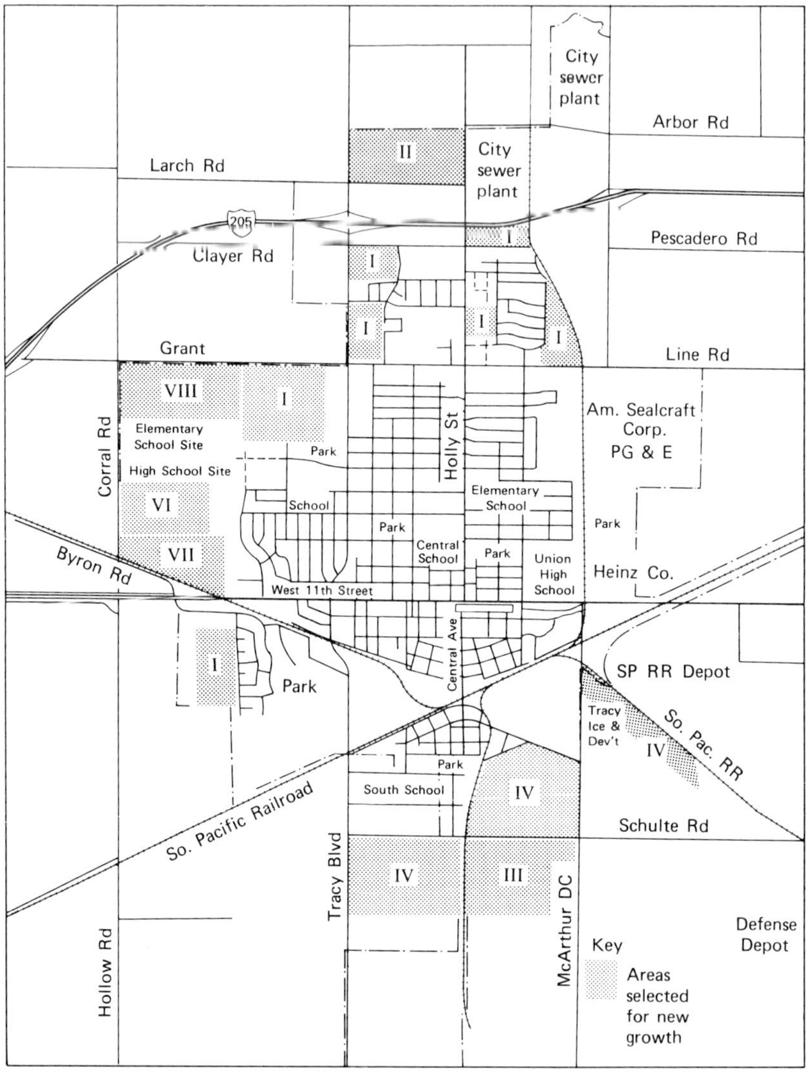

Figure 1

impact of these omissions is considerable; however, the point was made that new growth would pose important financial problems for Tracy.

At this point the Urban Growth Policy Discussion suggested a number of programs to minimize the effect of new growth on municipal finances and at the same time preserve the area's environment and agricultural production. The most important recommendation concerning agriculture and the environment involved the annexation of large farm areas to the city and their inclusion in Williamson Act Agricultural Preserves. After annexation these areas would serve as a limit to urban growth, forcing new growth to locate in relatively small areas adjacent to the city. To make this idea financially attractive to nearby growers, property taxes were to be financed on a district tax basis. This arrangement allowed exclusion of agricultural preserve areas, thereby substantially decreasing the amount of property tax the owners would be paying to the city. An added incentive was the provision of city fire and police protection (the cost was more than offset by the lower insurance rates after annexation). It was thought this combination of reduced city taxes and property tax stabilization through agricultural preserves would overcome the traditional reluctance of farmers to annex to cities and in the long run preserve agricultural production and prevent costly uncontrolled urban growth.

Other ideas developed in the study involved zoning changes to facilitate low-density residential development adjacent to the city and the elimination of a strip industrial corridor. The last change recommended in the study was an increase in city development fees to more clearly reflect the costs incurred by the city in providing services to new residential areas. Review of the cost-revenue figures clearly indicated that Tracy residents were subsidizing developers' profits.

Summarizing all the information developed by the Urban Growth Policy Discussion, the following recommendations were made to place new growth on a sounder financial and environmental basis:

That the city establish as policy an Urban Growth line for short-term growth. Within this area, the city will participate in the establishment of public services for urban use. The boundaries of this Urban Growth Policy line should approximate the boundaries of the eight growth phases, plus additional lands as indicated to allow for land sales competition to avoid inflation in raw land costs.

That the city establish a definitive Capital Improvement Program and Revenue Program to balance needs within newly developing areas. With the limited analysis to date, with the lack of programs for certain services, it appears on the surface that we have a net surplus of $995,239 by the end of the eight growth phases. However, as revenue lags behind costs by up to two years, we would estimate a deficit of $390,830 by the end of the eight growth phases. These figures will, however, be altered significantly when drainage, depreciation and overhead costs are applied and a larger deficit is anticipated.

That the city develop an updated Storm Drainage Plan applicable to the proposed development, as well as long-range needs.

That the city implement the detailed water plan for both immediate and long-range needs.

That review of water and sewer main extension policy be undertaken which would establish a more realistic cash flow basis for future growth.

That a basic service charge for development, including all costs and fees, be established so the developer can cost out a project at the onset and the city could program income from developers.

That the city undertake detailed investigation of the use of the California Land Conservation Act (Williamson Act) as a method of holding and reserving agricultural land for future urbanization.

That the city initiate the necessary procedures to amend the Community General Plan to reflect the proposed short-term land use proposal.

That the City Council request the assistance and cooperation of the County of San Joaquin in the elimination of the Interim Protected Agricultural Zone and establishment of appropriate zones as indicated by the land use plan.

That the city establish an Economic Development section within the Department of Community Development for the purpose of assisting and stimulating industrial and commercial development within the City of Tracy.

That the city undertake a cost accounting program to determine unit cost and depreciation schedules so that true cost can then be projected. That levels of acceptable service be adopted to enable projected manpower needs to be realistically evaluated and their costs anticipated.

Community Reaction

Publication of the Urban Growth Policy Discussion received wide publicity from *The Tracy Press*, the Chamber of Commerce, and other civic groups. The immediate reaction of the Planning Commission was to refuse to approve the proposed trailer park annexation. A number of city departments began discussing adoption of a capital improvements program and other accounting innovations. Beyond this immediate burst of activity, consideration of the study's recommendations has stalled. Reluctance to more seriously consider the study is attributable to two factors. Most important, the City Council in Tracy, as in other communities, hesitates to adopt any programs that will restrict its freedom of action. Recommendations in the Urban Growth Policy Discussion, the Council realized, require a number of definite policy commitments and difficult decisions. Once these are made, there can be no turning back, and the Council's conservative business orientation makes it difficult for members to see the need to regulate any sector of the local economy, including developers. For the most part the Council has adopted a wait-and-see attitude as of this writing.

Citizen reaction has also been slow in developing. This is not difficult to understand because most reaction is to some important activity that affects their lives, and thus far growth has not been terribly menacing to the average person. Compared with Livermore, the tax rate is stable, the schools uncrowded, and life pretty much the same as 10 years before.

To break this deadlock, the Planning Commission adopted the Urban Growth Policy Discussion and directed the Community Development Department to take the steps necessary to ensure implementation of its recommendations. With the City Council deadlocked over the matter, the department has taken the Urban Growth Policy Discussion to the people in an effort to continue consideration of the idea of controlling growth. The first group contacted was the city realtors, who were understandably nervous about the prospect of government interference in their field. Next the Chamber of Commerce, the Optimists, the PTA, and numerous other groups received presentations concerning the study. After this extensive period of discussion, the Citizens' Committee for Community Development, the citizens' advisory group to the Planning Commission, held public forums on the Urban Growth Policy Discussion in every part of the city.

During this public discussion an unusual alliance was formed in support of the Urban Growth Policy Discussion. Builders and realtors overcame their initial hostility to the plan when they realized that the ideas merely suggested rational growth, not a no-growth approach. Perhaps the builders and realtors feared that another bankrupt "Livermore" in Tracy would be harmful to everyone, especially their profession. Growers in the south Tracy area also began to support the study when a wave of vandalism convinced them of the need for the increased police protection they would receive after annexation.

Despite this increasing support, the council still has not adopted the study, but a number of recommendations have been implemented in the interim. The Chamber of Commerce has agreed to let the Community Development Department manage its economic development activities, thereby setting up a much more aggressive program to attract new industry. The necessary zoning changes have been made, and the county is considering developing more stringent agricultural zoning of areas adjacent to Tracy.

The key to the success of the Urban Growth Policy Discussion is its dollars-and-cents orientation. Most people can grasp the idea that new growth will cost money. As Will Rogers once said, "The way to wake a sleepy man is to pinch his pocketbook." In typical discussions on the study, it is not uncommon to see a man rise and declare his disgust that the town was going broke destroying "all that beautiful land." Such incidents demonstrate the advisability of using a financial-based study to bring about environmental awareness.

It is not contended that similar strategies will work in other communities, but the financial-based study has gone further than most expected in Tracy and may indeed lead to a period of rational, environmentally sound growth for the town. Perhaps as more cities pursue such studies the true costs of growth will be fully explained to each citizen, and the "growth budget" concept will be adopted to allow annual review of the cost associated with expansion.

Part VI

Information Systems and Modeling

Patricia A. Bonner
Lawrence R. Taylor

Mr. Taylor is a senior research planner, currently Chief of the Project Development Group of San Diego County's Office of Environmental Management.

An information system is simply an organized flow of information from sources to users. It orders the vast amount of available data into coherent pictures. It may use computers, but not necessarily. One of the key features of information systems is continuity. Information systems, by definition, are designed to provide data on a continuing basis, and the data content is restricted to known applications.

"Data banks," in contrast, are massive libraries of computerized data that may not have direct applications and may be updated infrequently to purge outdated information.

Information systems are built for two general reasons: (1) to aid decision making, and (2) to aid research. Decision making focuses on choosing optimum courses of action; research, in contrast, attempts to find new insights into natural and man-made processes.

At present, any objective observer can determine that most environmental data are mediocre, at best. Some environmental problems are well in hand; others are unknown. The decision makers appear to be overloaded with volumes of data, yet lacking other important information. Instead of solving the problems, environmentalists are interpreting mountains of data for certain subjects; for others they are searching for data.

Two major reasons seem to account for deficiencies in data: lack of data and data "channels," and the "channeling" of data.

An environmental information system, then, should first improve the environment by improving the quality of environmental data used in decision making. The system should strive to accomplish the following:

1. Generate greater regional consciousness (since most environmental problems cross jurisdictional lines).
2. Clarify objectives of public policy and action.
3. Provide measures of progress or retrogression.
4. Embody a reporting system that is policy and analysis oriented.

There are four main types of environmental information system.

1. Scientific—designed to transfer knowledge as opposed to data (essentially bibliographic).
2. Management—aimed at aiding the management process through project control and budgetary surveillance.
3. Monitoring—records the amount and nature of pollutants, pollutant damage, and physical features such as soils and wildlife.
4. Regulation—used to establish and maintain regulations.

Modeling uses information systems to simulate a "universe," its components, and their interrelationships. Models are becoming exceptionally popular in environmental management because of the growing need to have explainable interrelationships between variables rather than mere information about the variables.

Design of a model should be fully integrated with design of the data bases and other possible component systems to ensure maximum utility and acceptance. Before beginning to construct a model, ongoing models in the region should be examined, permitting an intelligent decision to be made about building independently or integrating. Basic questions should be considered prior to modeling efforts. What is the purpose of the model? For what question(s) are answers being sought? Can you identify the output(s)? Does the model provide the output necessary to answer the questions? What questions cannot be answered by the model? What will be the inputs? Are the inputs available? How much time and expertise are necessary to prepare the inputs for the model? How difficult is it to prepare alternative inputs for alternative simulations? Do the experts in the field consider the particular model(s) to be sensitive to real-world variables? What assumptions are built into the model? Does your area of study meet those assumptions? What input variables do you feel are of primary importance in your particular study? Does the model include these variables? How will the model relate to others in this region? What degree of precision is needed? How is location handled? How will output be displayed? What computer facilities are needed? What are the costs for a single run? Will the (money and manpower) costs of the model exceed the resources of the sponsoring organizations? Are the costs commensurate with the impact on decision making?

Building Environmental Systems in San Diego

LAWRENCE R. TAYLOR

In recent years, there has been considerable interest in the improvement of environmental decisions through the provision of better information. This chapter, which discusses some of the work done in San Diego, is neither a systems proposal nor a concept paper. Rather, it is an attempt to relay some painfully acquired experiences.

Background

Most of the environmental systems work in San Diego was begun under the Integrated Regional Environmental Management (IREM) project. However, IREM was able to learn from the comprehensive regional land use and transportation planning effort of CPO (Comprehensive Planning Organization), the local Council of Governments (COG)

Originally, the central concept of IREM was the development of an environmental information management system to help existing governmental authorities make environmentally related decisions by providing the authorities with environmental information at key decision points, and indirectly coordinating their activities by guiding their decisions across a set of regional goals. IREM was not to create another layer of government directly involved in managing very specific resources or directing specific urbanization activities.

The concept was not necessarily new, but it was considered to be a breakthrough in the effort to establish an effective management system. The system was expected to develop and manage information on a regional basis at decision points where potential existed for causing environmental change.

What are the decision points? There are many—too many. Governmental services in the San Diego County region are provided by the county, by 13 incorporated cities, and by nearly 100 independent special districts. The special districts supply single services to limited areas of the region. The cities are multipurpose and have exclusive land use jurisdiction. The county is considered to be a political subdivision of the state, responsible for the performance of such state functions as public welfare and air pollution control, although most of these functions operate throughout the region. The county also provides municipal-type services for the unincorporated territory.

In addition to the traditional special district city county form of local governmental organization, San Diego County has a Council of Governments—the Comprehensive Planning Organization (CPO). This organization provides regional liaison to the federal government for planning and metropolitan clearinghouse functions. CPO is also responsible for development of regional long-range land use and transportation plans.

Local land use planning is carried out by each of the cities and, for the unincorporated areas, by the county. Public works facilities are planned by the county, by the cities, and by the special districts. Major facilities, such as trunk sewers and freeways that involve federal financing, are reviewed by the Comprehensive Planning Organization. In addition, development pressures led the public to establish a special commission to control detrimental actions on the coast. Obviously there are numerous jurisdictions, all trying to optimize their operations. Such an optimizing process inevitably leads to conflict, competition, and subtle struggles for preeminence.

Identifiable environmental problems within the county are related closely to the overall phenomenon of population growth. One of the fundamental tasks is directing the pattern of land use, particularly of new housing needed to accommodate increasing population. Major parts of the region's chief consumable natural resource, its undeveloped land, are likely to be irreversibly committed to a long-term land use pattern without consideration of environmental effects beyond the boundaries of individual development projects.

Numerous other problems can also be traced to the clash of the natural environment with population growth, technology, and economic pressure. These include threats to wildlife habitats, air and water pollution, and uncertain urban amenities. Despite these problems, the overall environmental quality remains quite high.

The electorate in the San Diego County region is very conscious of the high quality of the region's physical environment. This concern may be illustrated by several items:

- The cleaning up of San Diego Bay through development of a metropolitan sewer system.
- On a smaller scale, the establishment of an experimental recycling system to treat sewage for use in recreation lakes in the community of Santee.

- Development of a major multiple-use recreational area in Mission Bay.
- Air pollution always ranks as the first or second source of concern in public opinion polls.
- The raising of the issue of growth per se in the last San Diego mayoral campaign, and the election of the candidate who favored redirecting growth, Pete Wilson.

Despite public concern and despite successes in several areas of the environment, the basic pressures still remain.

The situation facing IREM also structured its overall experimental program:

- It was attached to a governmental body having many bureaucratic channels.
- It was a short-term experiment, established with a life span of about two years. It was terminated in July 1973.
- The first period of the project suffered greatly from lack of structure. Considerable interorganizational damage was done before it was realized that a concern for the environment is no substitute for management.
- Expectations of results of the project varied widely. Some persons expected cold scientific analysis, others did not want "any more studies" and demanded advocacy. The early "nonmanagement" situation did nothing to clarify matters.
- IREM's broad mandate for environmental quality was been considered to be an implied threat by a variety of organizations, including certain governmental bodies.
- Several major, external, near-crisis demands provided IREM the opportunity to be "the right organization at the right time." These include the *Friends of Mammoth* decision on environmental impact reports, the rejection of the original air quality implementation plans, and demands for an improved environmental data base.

The responses of IREM staff to the situation just outlined were: (1) to run a series of specialized, pilot studies; (2) to assist other governmental bodies to establish ongoing, permanent functions; (3) to build a series of data bases and models, usable by a variety of agencies; and (4) to run the gamut of organizational, political, and personnel difficulties. Rather than enumerating all these experiences, let us discuss several of the more systems-oriented pilot projects and summarize lessons from these and other efforts.

Pilot Effort: Clean Air Project

Without a doubt, the largest single analytic effort in IREM was the San Diego Clean Air Project. This effort comprised about one-third of the total funding. In April 1972 the County of San Diego was awarded a $450,000 grant from the U. S. Environmental Protection Agency (EPA) to analyze alternative land use and transportation strategies for meeting air quality standards.* A matching contribution of approximately $25,000 was received from local sources.

Work was performed by IREM and consultant personnel, primarily the Rand Corporation of Santa Monica. The major product of the project was recommendations for improving air quality in the San Diego region. The recommendations were unique in that they were based on extremely sophisticated analysis techniques. Consequently, the resulting proposals were more than intuitive simplifications. Each possible means of improving air quality was carefully rated according to both cost and effectiveness. Cost-effective measures are not just direct payments; they involve many other impacts including noneconomic effects on socioeconomic groups.

Another product of the project was the development of new analytic tools that can be applied to other environmental concerns, such as water pollution, conservation of open space, and coastal protection.

IREM's Clean Air Project has employed a variety of analysis techniques to predict a wide range of impacts, including effects on air quality, regional economy, socioeconomic groups, and social costs of changes in transportation. In addition, the use of two dates (1975 and 1995) results in major complications in analysis. The 1975 analysis must be carried out in great detail, since 1975 is the year in which federal law requires "clean air." Millions of dollars and considerable social costs are involved in the decisions that must be made and implemented before 1975. On the other hand, 1995 represents the long-range aspirations of San Diegans. Since the Comprehensive Planning Organization is making most land use and transportation formulations for 1995, the IREM project is concentrating on the air quality for that date. The compatibility and possible tradeoffs between short-range actions and long-range goals are interesting aspects of the project.

The complexity of the situation has led to a variety of analytic techniques.

* Editors' Note: The results of the Clean Air Project have since been presented to the County Board of Supervisors, the California Air Resources Board, and the U.S. Environmental Protection Agency. Analysis indicated that for San Diego it would be more cost effective to place more stringent controls on fixed sources and require a major retrofit program than to dramatically increase mass transit. Project data served as the basis for a detailed transportation plan for the region prepared by county staff.

Following are some general observations that are useful in understanding these techniques.

1. Both 1995 alternative and 1975 "nominal" (existing trends) land use and transportation forecasts were prepared by the Comprehensive Planning Organization and the California Division of Highways, using models of urban development and transportation.

2. Possible policies for improving air quality were identified and tested with new models built especially for this project. These policies differ from the nominal cases just mentioned and include major transportation actions.

3. The 1995 alternatives have already been reduced to four major land use/transportation concepts. In contrast, the 1975 tactics are very numerous. A special screening process was used to reduce the number of 1975 tactics and combine them into strategies.

4. The promising strategies for 1975 were analyzed in depth in terms of impacts (transportation/mobility, air quality, costs, etc.).

5. The 1995 alternatives had their air quality impact predicted. Then the possible short-term (1975) decisions were reviewed within the long-range goals.

6. Comparison of the alternative policies was primarily in terms of cost effectiveness, including noneconomic costs. In addition, institutional impacts were considered.

The first stage in the Clean Air Project was screening, to identify the most promising tactics for the short-range period (1975 and 1977) from among the large number of possible actions. After screening, the selected tactics were combined into strategies that were analyzed in depth. The key to effective screening is to consider the individual tactics as part of a generic whole that in turn can be related to major issues. The final project report gives details on how the individual tactics relate to these major issues.Following are some of the major issues related to screening.

Will screening eliminate any tactics that may be promising on more detailed analysis?

Will screening satisfactorily deal with the short term within the long-range goals?

What are the tradeoffs between controls on fixed and mobile sources?

What are the tradeoffs between reduction of automobile driving and increasing controls on automobile engines?

Results of this rough screening process suggested that some of the more frequently discussed proposals for improving air quality may not be cost effective.

The second stage in the Clear Air Project was detailed analysis, designed to predict the impacts of the promising strategies (combinations of tactics) after the screening process. The detailed analysis included prediction of: (1) air quality at certain points in the region; (2) economic impacts, such as costs, employment, and possible requirements for subsidies or community taxes; (3) indications of transportation service, such as volumes and travel times; and (4) impacts on different socioeconomic groups.

The available transportation models were found to be unsuitable for environmental policy analysis. Yet transportation models are critical to the prediction of the impact on air quality of proposed changes in transportation. Consequently, a new set of transportation models was prepared. For similar reasons, a new model was built to enable the prediction of costs of bus systems when used in the heavy volumes necessary to improve air quality (as opposed to today's low usage rate).

Prediction of impacts on socioeconomic groups and the economy used standard methods. In these areas, the problem revolved more around finding the necessary data, a task that is often considerable. The use of these techniques was fully demonstrated through various examples, including the impacts of possible additional controls on automobile engines and on fixed sources.

The state of the art of prediction of ambient air quality is currently poor; however, it is rapidly improving. The problem is complicated in San Diego by the small number of monitoring stations (seven in 1973, six in the base year of 1970, in an area of 4300 square miles). The project employed a combination of older and relatively crude prediction techniques because of the emphasis on policy analysis, as well as restrictions by EPA. These techniques are "proportional" (i.e., air quality is directly proportional to emissions amounts) and "Larsen" (i.e., pollution levels over a year are similar to a log-normal mathematical curve). One innovation was the adjustment of the statistics of the Larsen technique to different measurement stations.

Finally, the detailed analysis included an investigation of the institutional constraints to improving air quality. The institutional analysis was warranted by the recognition that technical recommendations often go unheeded if they ignore the intricate relationships and jurisdictions of San Diego's organizations. For example, the county's Air Pollution Control District (APCD) regulates fixed sources. Land use is regulated by the 14 planning commissions (13 cities and the county), while the Comprehensive Planning Organization strongly influences future patterns of growth. Yet land uses *are* fixed sources. Does the APCD have the power to veto land use proposals that adversely affect air quality? Should it? Similarly, more than 30 separate agencies strongly influence San Diego's transportation. If the project is to be effective, these institutional strengths and constraints, however complex, must be recognized and utilized. In contrast to the mathematical models, there is no printed output from such staff work. Instead, progress must be measured by the quality of life in San Diego.

Pilot Effort: Regional Impact System

Concerning a separate but related set of topics, work has been progressing on a "regional impact system." Unlike the well-structured Clean Air Project, much of the work on the regional impact system has been done on an ad hoc basis, very loosely coordinated. Fortunately, workers have been able to draw on the experience of the modeling effort of the Comprehensive Planning Organization.

Since 1967, the San Diego Comprehensive Planning Organization has been developing models for regional land use and transportation planning. The models include a population and employment forecasting model, an urban development model (PLUM) and the traditional transportation models, plus mode split. A conservative estimate places their development costs at over $450,000 *not* including data collection. In general, the models have been effectively fulfilling the regional land use and transportation planning function. They have proved to be sensitive to the kinds of long-range policy choice that regional planners feel can be influenced, such as transportation accessibility and the supply and location of developable land. However, the classical land use/transportation models (as utilized in San Diego) were judged to be inadequate for environmental impact analysis.

The difficulties associated with using the land use/transportation models stem from the following conditions: (1) the logistics of running the battery are complex, time consuming, and relatively costly; (2) outputs are not in terms of environmental impacts; (3) the centralized nature of the models creates interorganizational problems. Logistics and cost have restricted the use of the models to "large" topics, such as new transit systems. The problem of limited environmental output was tackled by construction of a series of experimental, special-purpose, time-sharing-oriented data bases and models. Use of time sharing also solves the third problem. In other words, the large-scale land use/transportation models function effectively for regional planning work but are not satisfactory for daily environmental impact analysis. Consequently, they are being supplemented by smaller, special-purpose efforts termed the Regional Impact System.

The Regional Impact System includes the following components:

1. *Sensitive Environment Areas* (the Alert/Early Warning System)—the location of a proposed project is compared with the physical conditions already present. Outputs include geologic hazards, sensitive ecologic areas, historic and archaeological sites, high-quality vegetation, and soil limits.

2. *Air Quality*—the ambient air quality will be forecast based on emissions (including those derived from transportation) and a meteorology/air quality model.

3. *Government Costs/Revenues*—output includes all major costs and revenues to each jurisdiction. Attempts are being made to include indirect social costs.

4. *Resource Demands/Limits*—demands for resources such as water, energy, and land are being forecasted on the basis of per capita rates under varied assumptions. The definition of limits has just begun and will probably be keyed to spatial variations.

One major feature of the Regional Impact System is the use of a computer time-sharing network. Time-sharing networks are nothing new; the literature of information systems abounds with concepts and case studies. However, the effort in San Diego is different for three reasons:

1. No promises were made (in the hope that there would be no unfulfilled expectations.

2. There was no major investment in central computer facilities nor in the manpower required to support such equipment; instead, existing commercial time-sharing and teleprocessing services are being employed.

3. The primary impetus behind time sharing is *not* improved data processing efficiency but the securing of immediate access to data bases and models for a variety of independent organizations.

There is no doubt that many activities would be less costly in batch operations instead of time sharing. However, having batch operations by one group tends to raise organizational and psychological barriers to other groups. The use of time sharing is a deliberate experiment to see whether these barriers can be eliminated. If successful, the approach will mitigate some of the problems resulting from the fragmentation of decision making.

Conclusions

What can be learned from the IREM effort to bring strong analysis techniques to the environmental decision-making process? First of all, it must be recognized that many of the projects have not culminated. Therefore, many of the insights are still uncertain. Nevertheless, let us be brave and state them as absolutes:

1. In the competitive and critical world of governmental decision making, initial management actions set a tone that may dominate future work.

2. There seems to be a negative correlation between the permanency of the organization and the acceptance of its work.

3. The permanency of the organization is also related to the nature of the work program. The IREM Project funded a series of short-term, experimental projects. These projects have served to identify the appropriate methodology and have sometimes aided immediate decisions. However, these lessons must be backed by reasonably permanent funding; there appears to be no support for continuing systems.

4. The time-sharing network experiment appears to be a success. It has served as a gimmick to aid communication between a variety of organizations. A full-time resource person seems to be vital if there is to be effective interaction among groups. Such a person must be familiar with both the computer system and the projects of the various organizations.

5. Use of think tanks, such as Rand, offers the theoretical strength of universities with the timeliness of traditional consultants. However, it often requires dramatic adjustments in the operating procedures of both groups.

6. Recent literature on air pollution has presented methods for linking land use/transportation models to air quality models. Our experience is that performing the linkage is not as easy as stating the concept. For example, although land uses are fixed sources, land use data categories are structured for land planning, not simulation of emissions. Similarly, the battery of transportation models currently in use requires substantial care and feeding. These logistics problems prevent examination of a wide variety of options. Our response has been to build a new set of transportation models, models that are suitable for environmental analysis. Land use/transportation planning is the key to many environmental problems, but planning *tools* may not be directly applicable.

7. The concepts of tradeoffs and cost effectiveness seem to be well received by decision makers. The basic approach is understandable, although people sometimes get lost in the details or in concern for impacts on their own jurisdictions only. The ability to trace costs and benefits through different levels of effectiveness has been one of the major contributions of the Clean Air Project. Cost-effectiveness techniques promise to bring the environmental imperative out of the protagonist stage into more enlightened decision making.

Regional Modeling Studies

KENNETH E. F. WATT
WILLIAM BRYAN
ANTONE MIHANOVICH
YVONNE HUNTER

Overall Conceptual Framework

Regional modeling is, by its nature, a research task of considerable complexity, requiring teams of some size. The Interdisciplinary Systems Group group has several types of modeling efforts in progress. We are trying to build a library of submodels (transportation, land use, demography, pollution, transport, etc.), which can be combined into larger models that explore various policy options for alternative scenarios, presented to us by user groups. The submodels can also stand by themselves as a research and learning tool, as well as a means of conceptualizing the working system when the models are integrated. Also, global events affecting regional dynamics are being examined, and we have been modeling these and resulting effects.

Our overall modeling framework is contained in a gaming model known as "Calgame." Calgame is a highly aggregated mathematical model consisting of several key variables (e.g., airshed) at the state or regional level. It is utilized as a tool for evaluating the possible effects of large-scale political and technological alternatives on significant variables such as population, natural gas energy, other fossil fuel energy, forest resources, agricultural crop production, water and air quality, range land, municipal land, and government services. Dynamic in structure, the model is parameter-identified by using actual historical data on all variables and inputs to calibrate the interrelationships to approximate past trends.

Previous efforts have led to a continually evolving philosophy, recently so-

At the time of writing, all four authors were affiliated with the University of California at Davis.

lidifying into a point of view that now appears to have four principal characteristics. We itemize these, then explain each one in some detail.

1. The man–machine system of decision maker and computer simulation model and modeler should be regarded as continually evolving. Thus the system should be handled to facilitate the joint development of both the decision makers' knowledge of and the scientists' confidence in the system, and the realism and pedagogical usefulness of the models.

Our philosophy has evolved through several stages of man–machine interaction. Having come to the conclusion that decision makers have neither the time nor the inclination to regularly use regional models, we are concentrating on supplying expert opinion, on the basis of our model outputs, for alternative policy options, to problems posed to us by decision makers. The models usually take the form of simulating the effects of changes in certain key parameters. We thus avoid the complexity and controversy of determining the causal pathways by which a parameter can best be changed. We leave such decisions to the politicians, since we have no desire to usurp their authority or responsibility.

2. The discovery and mathematical description of key "driving mechanisms" in regions is the critical initial step in the development of our models. Sample hypotheses are: "age-specific birthrates in women are primarily determined by the age structure of the population," or "urban tax rates per capita are determined largely by urban population density, growth rates, and average distance between residential structures," or "demand per capita for gasoline is determined by urban area, density, modal split between public and private transportation, and gasoline prices." Such hypotheses are tested using statistical analyses of data, which then serve in developing mathematical statements of relationships between variables.

3. Because many members of the group are biologists trained in the temporal behavior of growth rates, we have always been acutely sensitive to the possibility that various diseconomies of scale may appear at high densities, population sizes, or growth rates. That is, growth rates should slow when a carrying capacity is reached, because of reluctance to change and human choice; however, growth often continues, and systems deterioration is the result.

4. This group has always recognized that economic growth of any given region is to a considerable extent influenced by factors operating at a global level. Three important factors external to each region are: global energy supply relative to demand, global food availability relative to demand, and the global level of air pollution. The first and second items will influence intraregional energy and food and land prices; the third will determine climate, hence crop yields.

Discussion

The Evolving Man–Machine System

We are convinced that no great technical achievement is involved in the construction of an extremely complex model by itself. Nor is any great insight incorporated in or revealed by vast assemblages of data alone. These two conclusions are quite obvious and will evoke little debate. However, a preponderance of the modeling effort completed in the past decade falls into one or the other of these classes and has had very little impact on real-world decision making. Originally we were convinced that the really large and challenging step was putting simulation models into an interactive format in which they are useful to decision makers, who would use them routinely as an aid in decision making on the basis of an accurate world view presented to them. (To this end, we developed a keyboard-with-video-display interactive systems that was maximally revealing of system behavior and placed minimal demands on the user.) This philosophical orientation mandates a model that is structurally simple and readily comprehensible. We have altered the original strategy of achieving direct interaction between the computer model itself and the decision maker. We are still interested in building simple and comprehensible models; however, the output needs to be presented to decision makers to provide a basis for quantified discussion of policy alternatives. Along with this presentation of results must be a discussion of why the decision maker should believe the results (reliability of the model). The model is then made available, via technical user manuals, to the staff of the decision maker, who can input their own data and run the model without need of our assistance. For example, our regional transportation model, now calibrated for the Sacramento area, could easily be adapted and used by staff members for other urban areas with similar population densities.

It is feasible to build a systems model, encoding the best existing knowledge on a given process, and to use this model to evaluate the effects of various policy options available to decision makers (as opposed to determining the "best" alternative for the solution of a given problem). Decisions will be made in any case, and in the absence of scientific input, they usually will be based on a simple intuitive model drawn from the decision maker's past experience. Thus the systems modeler can improve the decision-making process if he does no more than *accurately* produce a mathematical model based on expert opinion about the underlying structure and process, and the consequences of this structure for various policy options.

By having on hand submodels that are basic to all the specific problems of different user groups, we can select and combine the models from our library to build "larger" models to answer specific user questions about alternative policy options.

In many cases our models challenge belief systems because the conventional wisdom is typically expressed in the form of particular indices of systems that *appear* to be revealing but in fact give a *distorted* view of the state of the system. (To use a simple example: GNP per capita in constant dollars would be declining long before total GNP in current dollars.) Thus most people have a distorted view of cause-and-effect mechanisms. We hope to correct this state of affairs through careful selection of indices for use in describing systems states. A project now in progress uses a flip chart approach to present the difference between commonly held beliefs or viewpoints and observed statistical reality. A "faulty" assumption is present on the first card (e.g., pollution concentrations will never be severe enough to have a significant impact on death rates). Two graphs follow, one describing what the observed data would reveal if the assumption were valid, and the other the results if the assumption were invalid. A third graph indicates what the data *do* indeed reveal, supported by a concluding statement (e.g., pollution concentrations have a significant impact on death rates). A consistent aim of the group is to express everything in terms that clarify the state of events in the system.

Driving Mechanisms

By "driving mechanisms" we mean any process that has a significant impact on critical state variables in a region (rate of population growth, rate of conversion of agricultural land to other land uses, oxidant concentration in the air, etc.).

A good example of how we use driving mechanisms is age-specific birthrates in women, one of the two most important vectors of variables in determining future population trends. (The other is age-specific net migration rates.)

We can project birthrates, year after year, in either of two ways. On the one hand, we can explore the consequences through time, for the region, of age-specific birthrates changing over the next few decades in some predetermined pattern. This pattern of temporal change would be specified for the computer in the input data net for each run. On the other hand, the model itself can calculate realistic age-specific birthrates for each year, using certain data from the previous year, and the equations describing the driving mechanisms for determination of birthrates. Clearly, the computer will generate scenarios under either system, and either system can expose the effects of particular vectors of birthrates. But the second approach, using the additional driving mechanism equations, is clearly the more realistic in that the computer simulation operates as a closed system where each variable is determined only by the prior state of the system.

Diseconomies of Scale

Several types of diseconomy of scale are being explored. For example, preliminary statistical analysis shows that there are no clear-cut economies of scale as city size increases. Results indicate that per capita operating costs increase rather than decrease with the size of a city. In many cases this increase in per capita costs is coupled with a decrease in the quality of services (e.g., fewer acres of recreation land per capita or higher crime rates).

We are becoming increasingly interested in the dynamics of reaching urban or regional saturation levels or diseconomies of scale, and of the internal or cultural forces that either promote or prevent environmental saturation and resource depletion. Many of the results of our models have the offshoot aim of educating the public to these diseconomies and carrying capacity levels. For example, a common contention is that more and bigger freeways relieve congestion and increase traffic flow. This is not always correct, and often the enlargement of a freeway only acts to reduce traffic flow rates. Frequently freeways stimulate more growth and more automobiles, while at the same time acting to increase pollution levels and energy consumption. Thus our example of the effects of a freeway challenges prevalent belief systems and attempts to reeducate the public against perpetuating environmentally dangerous growth patterns.

Global Impacts on Regions

The international system and economy is an extremely interdependent system, in which events in one place have consequences almost everywhere else. As domestic fossil fuel shortages worsen, more fuel will have to be imported, and more food will have to be grown and exported to pay for this imported fuel.

As the need for increased American agricultural production to meet domestic needs and foreign commitments becomes more evident, agricultural land will become more valuable. An increase in production due to advanced technology will definitely result in some greater yields; however, our research has shown that fertilizer applications, for example, only increase yield to a plateau—thus continued fertilizer use is not the answer. On a regional level, this finding will have an impact on zoning and development procedures to meet the increased demand for agricultural products and land.

We are acutely aware that global pollution levels *can* and *do* affect weather, which in turn affects agriculture. However, our efforts at modeling the link between global pollution, global weather, and regional weather have been thwarted by lack of good data, and we have terminated this exploration. We have redirected our work in this area to regional airsheds, with the hope of obtaining more accurate data.

Tasks Required to Construct Models

Data Collection and Organization

Before any analysis of driving mechanisms or model construction can begin, it is necessary to assemble the data. On many jobs, we put ten times more effort into this step than into all the work on analysis and modeling combined. Some explanation of the difficulties of this process is offered below.

Immense problems are encountered simply in finding data. County and federal governments are often unaware of the data being collected by each other's agencies. Also, in many instances critically important types of data have not been perceived as necessary by any level of government and have never been collected. In general, there is only scanty information available on the rate at which land is converted from one land use category to another, per person added to the population, and measures that have been published are often very controversial. Many important data are published long after they would be useful in decision making: air pollution mortality data and natural fuel importation statistics are examples.

Many types of data that would be of great use in regional studies are hard to come by. For example, it would be helpful to know, for each year and each city, the number of revenue passenger miles of transportation delivered per unit of fuel consumed, for each of a variety of transportation systems, and also the total fuel consumed in each type of transportation per person in each year. In fact, very few of these statistics are available. There is little information on gasoline consumption by taxis. There is immense controversy about the number of revenue passenger seat miles of transportation delivered per gallon of fuel in commuter trains and intercity trains. Fuel efficiencies for novel transit systems are simply not known, yet such figures should be a key element in future planning. Where such data are available, there may be difficulties in pooling or merging different types of data because the aerial base for measurement varies. Data may be collected for the city, county, standard metropolitan statistical area, or urbanized area. The demand for fuel in a region will be strongly influenced by the degree of interspersion of places of work, homes, and stores; yet measures of this variable are not available. Compatible data for city expenditures are difficult to come by, since different means of collecting, recording, and storing are used, not to mention multiple means of obtaining financing.

Statistical Analysis Techniques

Our data analyses begin with the invention of hypotheses to test. To test the hypotheses we use three methods. For some relationships, we conduct statistical analyses of the way one or more variables influence a dependent variable through time (longitudinal analysis). Research on the determination of birthrates by the age structure of the population has taken this form. In other cases, where the data are available, we analyze the way a variable changes from place to place in a particular year in association with changes in other variables (cross-sectional analysis). For example, to find the effect of price and city structure on gasoline consumption per capita, we used nonlinear regression analysis to account for differences between 88 standard metropolitan statistical areas in gasoline consumption per capita in 1967. In a few cases, enough data are available to allow for analysis of place to place and year to year changes jointly. For example, air pollution concentrations are available for each of many different sites, and for many years at each site. Age-specific mortality rates for corresponding site–year combinations may be calculated from census bureau population estimates, and unpublished state data on numbers of deaths.

Since preliminary analysis of relationships use tabulation and graphing of families of curves, we can explore the way in which two or more independent variables interact to influence an independent variable. We infer the form of the mathematical relationship from the graphs, test our inference using transformation of variables and mathematical graph paper, and finally obtain the best parameter values using regression analysis.

The systems model, in turn, is developed on the basis of flow charting of the regional system. The flow charts indicate the nature of the causal pathways that drive the regional system and guarantee that important variables and causal pathways will not be overlooked. To pass the acid test, of course, the systems model must run and generate sequences of data that match the real world, insofar as our experience and the experience of other experts can determine.

Example Models

Two models are briefly presented as examples of how the previously described concepts were utilized in the solution of real problems.

Transportation Model

The primary objectives of a transportation study were to analyze the alternative transportation options available in an urban fringe area, and to quantify the impacts of implementing these options. The options to be analyzed consisted of a series of strategy or policy decisions faced by decision makers with respect to regional transportation. The options included such items as: moderate improvements in mass transit service, equalization of mass transit and private auto commute travel times, doubling of costs associated with private auto usage, and staggering work start times of the employers in the area.

A computer-based model was developed to simulate the area's transportation system. To utilize the model, a strategy of interest plus other relevant information is input to the model. The model, using various established relationships, performs calculations and presents outputs that reflect the effects of the input strategy. It is thus possible to compare the effects of implementing the various alternate strategies or policies by decision makers.

Model Inputs. The basic model inputs consist of the population level of the region—either current or projected—the strategy/policy of interest, technology in terms of air pollution and fuel consumption rates, and base year attribute values such as modal split and auto occupany rates. These inputs form the basis of calculations performed by the various submodels, which together makeup the overall model.

Submodels. The relationships of the submodels and the data inputs to the model are illustrated in Figure 1.

Based on the population inputs, the person-trips generator submodel calculates the number of trips expected in the categories home-to-work and "other" trips. The modal split and auto occupancy submodels allocate the person-trips generated into mass transit trips and private auto trips, including number of persons per auto. The allocation is performed by means of an elasticities approach measuring the change in modal split and auto occupancy (from base year values) as a function of their sensitivity to changes in cost or travel time.

The corridor allocation and travel submodel assigns the various person-trips to various traffic corridors within the region to simulate peak-hour traffic demands on arterials within the area. This traffic allocation is based on such considerations as average vehicle speeds, average travel distances (both work and nonwork trips), and assumptions concerning type of highway used (freeway vs. nonfreeway).

The energy use and the emissions submodel, and the transit use submodel, based on the outputs of the other submodels, calculates the total energy utilization and pollutant emissions for all transportation modes. Energy utilization is based on assumed miles-per-gallon values for autos and buses. The emissions calculation considers such factors as the age distribution of the automobile populations, average vehicle miles traveled per age class, and vehicle speeds per trip type.

Model Outputs. The model outputs include the following items: peak-hour traffic volumes at preselected locations, total transportation energy use, resulting modal split, resulting auto occupancy rate, and peak-level pollution

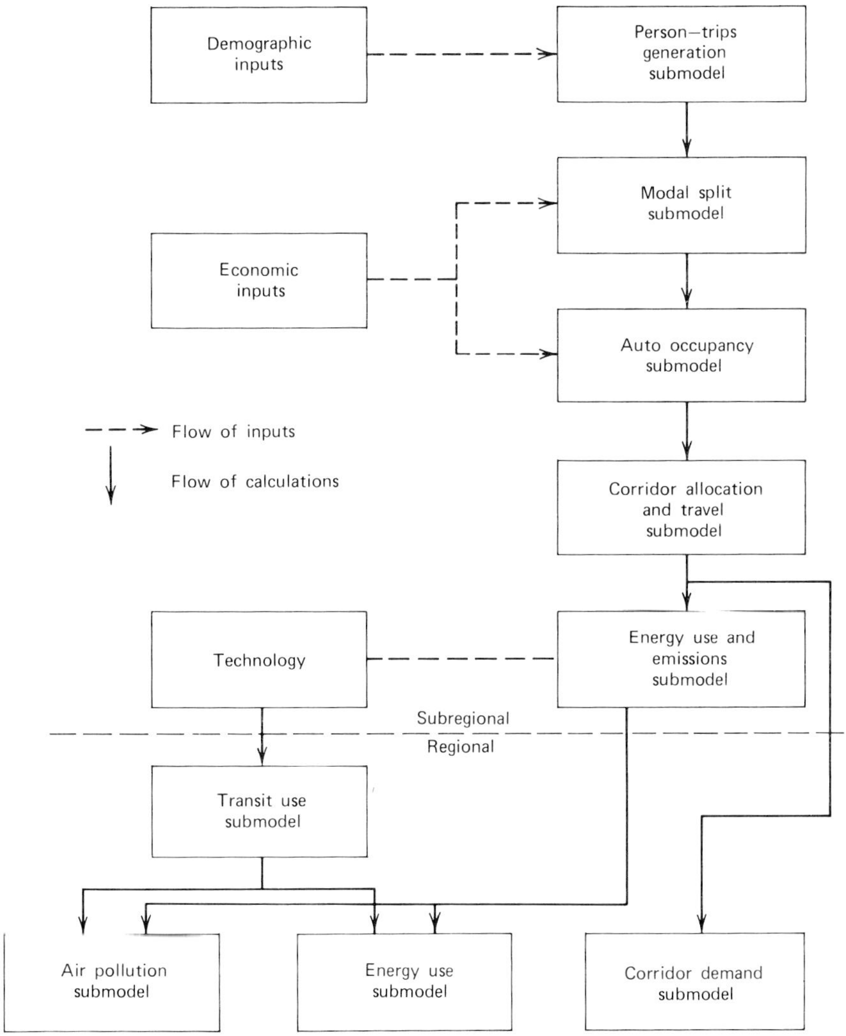

Figure 1. Regional model submodel composition and program flow.

levels, including the probabilities of meeting specified air quality standards. These outputs can be developed for various assumed scenerios resulting from various transportation policies and strategies. Thus the transportation model is a policy-oriented tool. Its greatest utility lies in its ability to quickly compare the impacts of a variety of scenarios on regional air quality, energy use, and congestion on selected routes.

Table 1. Sample of the Output

	Base Year	Base year + 5 Years	Base Year + 10 Years	Base Year + 15 Years
Item[a]	Initial Values	Percentage Change		Final Values
Population	1371342	5.14	6.34	1059087
Residential land	72360	6.53	4.52	84086
School land	5790	−7.20	1.43	6288
Recreational land	13192	5.13	6.34	15966
Auto emissions	445365	6.32	5.48	533023
Water	236613	8.76	9.29	313681
Natural gas	29276	18.07	14.50	42957
Electricity	7156485	31.41	27.60	15157171

[a] Land is in acres, auto emissions are in 1000 pounds, water is in 1000 gallons, natural gas is in 10^6 kW.hr, and electricity is in 1000 kW.hr. Base year is 1971.

Regional Resource and Urban Planning Model

A second type of model is mathematical; it projects the effects of various types of growth policy on the basic resources of a region. This model is useful for studying such problems as:

- If a population increase of x percent per year is projected for a certain region for a 10-year period, what will be the annual energy requirements to support this growing population?
- If a regional planning department, in an attempt to reduce urban sprawl, limits the land available for housing, what will be the long-term effect on housing mix (multiple vs. single-family housing units)?

The basis of this type of model is a series of relationships between population and land use, energy, water, and transportation requirements. Then as the various demographic values are input to the model, calculations are performed that project the resource demands over time. An example of the output is presented in Table 1.

A second example output relates to the use of the model to study the effects on housing mix of limitations on land allocated for housing. A simplified block diagram of the model and example output appear in Figure 2.

Part VII

Role of Special Interest Groups and Industry

Patricia A. Bonner

Government at any level can and will exert limited efforts to clean up and maintain environmental quality. Legislation can be passed and standards imposed at a certain speed and in limited numbers. Enforcement is expensive and time-consuming. Priorities are set to reflect governments' understanding of problems and needs; without citizen input, nothing more will be considered.

Individuals, community organizations, and industry owe it to themselves, their progeny, and their communities to make their views known. Representation at governmental meetings, petitions, formal policy papers, media interviews, articles, and associations aimed to achieve environmental objectives all are useful.

Government usually reacts to numbers and consistent presence or pressure, to organized factual presentations without emotional overkill, and to sustained representative interest. Only an organized citizen group, a retained lobbying group, or an industrial interest can provide the manpower to do the fact-finding work, the attending, and the funding required to mount such an effort.

Such groups can do more than pressure government to (re)arrange priorities, (re)structure programs, and pass legislation. Industry can set its own high standards to meet, bettering those of the government. It can work with community groups to sponsor beautification or recycling projects, organize seminars and workshops for schools, companies, or citizens, and publish environmentally oriented (energy-saving, air-cleaning, noise-reducing) suggestions for home and/or business. Industry can build its new facilities to conserve energy, design its processes to reduce industrial waste, utilize paper wisely, aim research and development into ecologically sound areas, and work to promote economically feasible environmental legislation.

Citizen groups can perform many of the same types of activity, but are frequently hampered by funding limitations. Most depend on volunteer research, supplies, printing, media help, voluntary attendance at hearings, and so on. Their job is harder to accomplish than industry's. It is sometimes even more difficult because so many single-purpose groups grow, emotionalize the issues, and dissolve; sometimes after successfully championing a cause; sometimes because money or enthusiasm has dwindled. Such groups serve their purposes when only a short-term single thrust is needed; for example, to block filling a canyon, destroying a park, or opening a gravel pit. However, over long periods only those groups which can sustain interest and maintain membership can retain credibility. The Sierra Club, the Izaak Walton League, and the League of Women Voters have done so on both local and national scales. In most communities a long-established, locally strong, environmentally concerned organization grew out of a conservation/beautification organization that began in the 1950s or early 1960s. Community groups today are seeing the need for coordinating environmental action and pooling facts and membership power to work for priority problems' solutions.

Successful initiatives are being carried out in states across the nation. Citizens are supporting environmental legislation and candidates who have stated their environmental interests. Industrial firms are working toward cleaning up the environment and with legislators to pass environmental laws that accomplish cleanup without unnecessary economic impact.

One major problem remains. Although environmentalist citizens and the representatives of industry work with government, they do not meet with one another. Often when they meet, it is to confront the "other side"—the "eco-freak" vs the "profiteer"—rather than to rationally discuss their views. To fully achieve their respective roles in environmental management, special interest groups and industry must find common ground and work cooperatively in and out of government at all levels.

The Price of a New Home

WILLIAM H. KRONBERGER, JR.

Contrary to popular belief, the builder is not unaware of the environment. A myriad of reasons exist to promote his growing awareness, including the obvious punches to the midsection recently taken in the form of *Friends of Mammoth*, coastal commissions, and other local actions affecting his pursuit of business. Like jackasses, I suppose, sometimes you have to hit them between the eyes with a 2 × 4 to get their attention. Let me assure you, and everyone else, that the environmentalists have the builders' attention.

I do not propose to discuss the question of environmental protection in terms of how the builder feels about it or how wonderful his new awareness is. Rather, I will undertake a course of discussion those in the stereotyping business would have smugly predicted: a few questions of economic reality. I do this not to build any groundswell of sympathy for builders or to predict economic catastrophes associated with unemployment and alleged economic losses to individual landowners associated with environmental action; instead, I want to emphasize that our national policies relative to the environment and to housing are not on the same course. There is a conflict between these two national priorities that needs resolution before we can settle down to the job of providing every American with a decent home in a decent environment. This conflict should not be used as a reason for slowing down environmental protection on the premise that the two policies are mutually exclusive.

Costs Outline

I propose to treat the question of the environmental price tag on new homes in two phases. One phase I shall term "tangible costs," for want of a better label, and the other "intangible costs," lumping the various fees included in the price of a home which I consider to be "environmental" under the heading of "tangible costs." The pain, frustration, and sizable dollar investment required to

Mr. Kronberger was the General Counsel for San Diego's Building Contractors' Association; he is now in private practice.

feed oneself into the only known perpetual motion machine—the planning process—become the "intangible costs." The two "costs" are relevant, since both are allegedly designed to produce a better living environment for our citizens and both are, in fact, paid for by the home-buying American public. Any illusions to the contrary (e.g., that the builder is picking up the tab), ought to be eliminated at the outset.

Builders, in a sense, reflect society's image of itself. Builders build what the public wants, and at the prices people are willing to pay for the work. Builders do not cause population growth and the demand for new housing. They are merely a tool of demand satisfaction. No one in his right mind will build houses that people will not pay for. Today environmental quality is a high-priority item. Everyone agrees on the desirability of environmental protection programs, but what people are willing to pay has not yet been ascertained. I hope here to outline a few of the costs as they now exist. There may be some who will not consider these costs to be "true environmental costs," yet they are a fact of life and to my layman's way of thinking reflect a concern with broad environmental quality.

Tangible Environmental Costs

I would like to begin discussing tangible environmental costs by borrowing a few lines from economist Milton Friedman of the University of Chicago. His comments, believe it or not, can be found in the February 1973 *Playboy* interview. To paraphrase Dr. Friedman:

> Even the most ardent environmentalist does not really want to stop pollution. If he thinks about it and does not just talk about it, he wants to have the *right* amount of pollution. We cannot really afford to eliminate it. The answer is to allow only pollution that is worth what it costs. . . . The problem is to make sure that people bear the costs for which they are responsible. . . . There is a romantic notion that by cracking down on the producers, we will somehow end pollution without any increase in prices. Nonsense. Why should not consumers bear the increased costs? They themselves are the only real producers of pollution.

So it is in the housing industry today. Rightly or wrongly, a large number of decision makers, and the public themselves, have decided that the "new guy" is the "housing polluter," as it were. Since he is the cause of the problem, a point with which I heartily disagree but will not stay to debate, he should bear the costs. Some people react to the problem of the "new guy" by actions to "stop or slow growth." This approach assumes much in the way of permissible control, and I firmly believe that our region will continue to need more housing, whatever growth policies are adopted. My reasons for this belief are many, including in-migration, births, new family formations, and family dissolutions. I suppose I could sum it up with a simplification—we will continue to have population growth as long as sex is more popular than dying.

What are some of the environmental costs now being placed on the polluting home buyer? What costs do new home buyers pay *in addition* to the taxes they share with their neighbors, and in many cases, with their parents across town? I emphasize the term "in addition" because it is important to remember that new home owners pay the same taxes as everyone else.

As a starter, the new home buyer contributes what is known as a park fee. This cost varies throughout the state and depends on whether the home builder supplied the local government with dedicated park land or actual cash. For purposes of our discussion, the County of San Diego's fee of $200 per dwelling unit will suffice. That fee is, of course, part of the sales price.

In addition to this park fee, we may add the relatively new school fee. Although not authorized anywhere in the law of the state, and specifically declared illegal in every court case testing it, local governments are exacting such a set fee from new home buyers to supply educational services. That these new home buyers continue to service the debt on existing educational facilities throughout the district is of no consequence. The cost of educational environment, *in addition* to property taxes, is approximately $750 per dwelling unit in our region. This polluter's tax, the price for bringing children into a community, is again tacked on to the price of the new home, even when the home buyer has no children. Thus far, our polluting home buyer has paid $950 in fees as the price for converting raw land to use as shelter.

We should probably add the costs of using the home buyer's own personal polluter, the automobile. For our purposes here, it would tend to overstate the environmental costs if we included the costs associated with providing streets within a project made necessary to serve the traffic stimulated by the new development. I do not, however, think it would be out of line to include the amount of money added into the new home's sales price attributable to construction of major thoroughfares used by other residents of the community to travel from one point to another. Here the new home buyer pays his fee for the benefit of the community's automobile operators as a whole and, in that sense, bears an extra burden over and above that caused by the project he has moved to. In the City of San Diego, council policy 200-1 places the cost for major thoroughfares on the new home buyer: it works out to about $250 per home and brings our new home buyer's costs up to $1200.

Finally, we cannot overlook the California Environmental Quality Act (CEQA) of 1970, as amended. On September 21, 1972, the California Supreme Court, in its now famous *Friends of Mammoth* decision applied CEQA to private projects requiring submission of Environmental Impact Reports (EIRs). These reports are now a fact of life, and they are expensive. The cost is difficult to calculate because it varies with each project. For purposes of illustration only, however, let us assume a 100-unit project with a $10,000 EIR. From what our members tell me, that is not terribly unrealistic, and it means another $100 to add to our fee schedule, increasing the cost of a new home by a total of $1300.

There are, of course, some imperfections in my figures. I do not present them as constants, but only as examples of the kinds of "environmental tax," if you will, that the new home buyer pays. These are tangible. They exist, and they are easy to identify. For want of space, I cannot go into the many side costs involved in the recent reordering of social priorities to environmental protection. I cannot, for example, spell out the costs in terms of sales prices for grading ordinances which in essence cut down on usable land. Nor can I figure per house costs associated with increasing required parking spaces per dwelling unit or increased setback and minimum yard and landscaping requirements. The exact per unit cost to hook up to sewer systems attributable to increased treatment standards is impossible to isolate. Likewise, the inspection and enforcement costs associated with policing these new regulations which find their way into hearing fees, application fees, building permit fees, and so on, is incapable of fair allocation. However, these costs *are* going up. I feel confident that the total would at least equal the cost of the four fees I outlined. I think I have made the point that right, wrong, or indifferent, environmental concern finds its way back to the product in the form of higher prices. In a nutshell, Americans will pay more for less home in the future, with increased environmental quality taking the place of the extra room or patio.

I do not quarrel with the wisdom of passing along these costs to consumers. Nor do I question their need. I suppose these costs are minimal when compared to providing a decent environment. At least, they are minimal if you are middle class. I am not so sure, however, when we start talking about shelter for lower-income families. At that point, our two national priorities of a better environment and a decent home for every American run smack into each other.

Since the crying need for housing is a fact of life, houses not built, or houses whose higher costs are charged to the environment, means less housing for those who need it most. And the world being what it is, environmental protection is likely to be paid for out of the hides of the poor. I do not say that housing for poor people and environmental quality are mutually exclusive. We can have both, but we need to reconcile the two policies first.

Intangible Environmental Costs

Let us now move from the area I have labeled tangible environmental costs to my second round of costs, entitled intangible and related to the system itself. The heart of any discussion about the environment is the planning process. It would be futile to consider the environment without relating it to the process by which we reach land use decisions.

It is probably in this realm that the builder reacts most negatively. Although it is difficult for a builder to agree with the rationale that designates the new home buyer to pay a greater share of environmental costs than his equally polluting neighbor, he can at least pass the cost along fairly easily. Coping with the system is, however, a different problem. It is like approaching Jello; it moves, but you are not sure whether it is alive and whether its shape will change according to the heat.

Most builders consider themselves honorable men. They only wish to satisfy their customers' demands as efficiently as possible while obeying the rules. However, it is easier to make that statement than to act accordingly.

There are two basic defects yielding higher costs in the present environmental planning process: (1) the process is a patchwork of layered regulation with costly, time-consuming duplication, and (2) at no time are decisions final. There is, in reality, no commitment to a builder of project approval until it's actually built.

Let us first face the issue of layered regulation. One year ago the builder thought he was caught in a maze of repetitive public hearings. He looked on local government with a certain degree of confusion, wondering how he should approach this giant structure with multilevel approval points. On one hand he was told to be creative, deviate from the normal cookie-cutter tract. On the other hand he could tell from assessing the system that he would have to pay dearly in terms of time and money to deviate. Sometimes it is easier to conform to the standards than to enter the maze.

An example would be the fellow who could build duplexes all over his property but wants to include open spaces, green belts, and recreation centers in a cluster development. Chances are, he would have to go through a series of Planned Unit Development type hearings to do so. Again, sometimes it is easier to build the duplexes. The problem is that all too often government has responded to the situation not by streamlining the system but by making it harder to build the duplexes, thereby compounding the problem.

That matter was a problem one year ago. Today we have an even greater problem of repetitive duplication mixed in a sea of uncertainty. I refer to the application of the Environmental Quality Act of 1970 to private development and the creation of regional coastal commissions by Proposition 20 in November 1972. The duplication involved in these two environmental efforts ought to be obvious to all.

A glaring deficiency comes to mind with respect to CEQA: it is a procedural act, not an environmental act. It sets up only another process a builder must go through before getting permission to build. The heart of CEQA is the compilation of a list of the project's environmental effects. Exactly what happens after this laundry list is prepared is the same as the earlier result: nothing. The antidevelopment people and the builders will be addressing their local councils with the same kinds of pitch they were making before. The only addition will be facts, and sometimes I have to concede that the old adage, "don't confuse me with the facts," never seemed more appropriate.

The act does nothing toward resolving the basic land use questions. If land is, for public purposes, not suitable for development, what happens to the land owners? Do we classify them, as many do, as "speculators"? It should be obvious that many of these "speculators" are middle-class, elderly people who have invested their life savings in parcels of property, in natural reliance on an established system of laws, customs, and practices. Maybe such investment is no longer practical, but can we write off those investors with a single stroke as antisocial? Should they not receive equitable treatment in light of practical political expediency? Is it not this failure to treat the question of compensation along with environmental protection that has so often led to frustrated legislation? And when we do get legislation, is it not nearly always procedural, such as CEQA? The biggest single issue in environmental protection is probably *density*, yet what does CEQA add to that? Nothing.

Even though a builder may have submitted his EIR for a rezoning, the environmental issues are *not* put to rest. He still has the hurdles of the Subdivision Map Act, Planned Unit Development ordinances, and a maze of often contradictory public hearings. After this monumental investment of time *and* money, he may have project approval.

What has this long road brought the builder? In some cases the right to appear before another layer of government, the Coastal Commission. The commission is then charged to consider most of the materials involved in the EIR process, but unlike CEQA, the Coastal Commissions involve a substantive decision-making process, treating the environmental issues squarely.

I do not wish to go into the utter choas in the building industry caused by Proposition 20 because of the bureaucracy's inability to administer it; rather, I would ask, Is not this putting the cart before the horse? To get a substantive environmental decision—still with no plans for compensation to injured parties—you have to go through all the procedural hoops of CEQA and project design, and the host of technicalities that go with processing maps through local government.

By the time a builder gets to the Coastal Commission, he has likely spent many dollars in design and coping with the system. These monies should never be spent before basic decisions are made, if we were truly interested in a responsive decision-making process. By virtue of the multilayered structure in which we operate, at no time are decisions final. Requiring investment of many dollars to navigate the labyrinth of rules leading up to the Coastal Commission is like buying a ticket to your own hanging. Going through all the steps entitles you to reach a point where you may be shot down.

Increased Costs Summary

I have now touched on two aspects of costs involved with environmental protection. We have briefly treated the cold, concrete identifiable costs of parks, schools, thoroughfares, and EIRs, reaching the conclusion that the polluting new home owner pays at least $1300 in surcharges in the price of his new home. The intangible costs related to other fees and dealing with a system not designed to reach a decision until the very end of a long procedural or deal have been discussed but are much harder to assess with a dollar price tag. A conservative estimate would double my $1300 figure. I suppose the question now is, So what?

Is "so what?" our national priorities conflict? To create a better environment, we tax the new home owner. On the other hand, we are committed to homes for our citizens, to give them a piece of America, if you please. Unfortunately, the former policy has the effect of making the latter policy unachievable. We cannot increase costs on one hand and make lower priced housing available on the other.

Let me give a concrete example of pricing the poor out of the housing market. The lowest priced home in Mira Mesa in October 1969 was $14,995. That was a single-family dwelling with a single-car garage. Since 1969 we have added park fees, school fees, road costs, doubled the costs of sewer and water hookups, and increased builders' overhead because of interminable delays. The effect: the same house today sells for $19,500. I do not pretend that *all* the increase is due to environmental policies. I do believe more than two-thirds of it is. The point is, and the sales statistics bear it out, that we have priced a large segment of San Diego's population out of the housing market. Sales surveys in 1969 and early 1970 indicated a balanced racial and economic cross section of purchasers in Mira Mesa. Such a cross section is no longer found. The City of San Diego recently enacted an ordinance requiring additional parking capacity. Thus in March 1973 the house I have just described was taken off the market, leaving $22,500 as the rock bottom housing price.

Most of us in the room consider $22,500 a very modest amount to pay for a house. That is probably because we can afford $22,500. Large numbers cannot! To them, $22,500 is an astronomical sum.

Middle-class home buyers may be able to pay the increased costs of environmental protection. They may even desire to pay them. The unfortunate part is that the poor cannot. Most experts agree that the provision of decent housing in suitable neighborhoods in our metropolitan areas for *all* classes of citizens represents the primary social need of our country today. We are not performing!

We need to preserve our environment, but at the same time we need certain types of development. However, there is a tendency to put the environment in a paramount procedural position in *every* case. No other element of a general plan must be specially found in each instance as a precondition to local action, nor does any other objective incorporated in a general plan structure require a specific impact report. Those persons interested in solving our neighborhood problems, in cutting the white suburban noose, and in finding housing for moderate- and low-income families might well ask: why should not each proposal for a park or for open space be considered in the light of a low-cost-housing impact statement? Why do the Attorney General and prominent legislators show such energy on behalf of environmental preservation, which essentially serves the rich, while the housing element provisions of our planning act, which might provide some relief for the urban poor, are evaded with impunity?

A clean environment and housing for the poor are both attainable goals. If we do not find a way to minimize the impact of environmental costs on the low-income-housing scene, we may inadvertently trigger a damaging conflict between environmentalists and the poor. We must recognize the problem now before the hint of such a situation arises. We must reconcile the impact of environmental costs on low-income housing at the outset, so that everyone can work for a better environment.

Considerations

So much for berating the system and casting invective toward the bureaucrats. I do not propose to have *the* answer. I do feel, however, that a few matters need consideration.

1. The conflict between environmental policies and housing goals must be addressed if we really want to provide a decent home for every American. Maybe we even want to adopt a sliding scale of environmental necessity based on home buyer income. I do not know. I do know that if we fail to relate environmental costs to the need for low-income housing, the future holds serious problems.
2. The system is straining to the breaking point to absorb all the procedural patches added. We must institute a system that treats the questions of land use and compensation or face a continuation of the layering of our decision-making process.
3. We must revamp our system to make environmental decisions at the very earliest point. This involves recognition by all parties that density plays a key role. Perhaps a system of density zoning would go a long way toward clearing up the procedural maze. Such a system would involve a density decision at the first stage following public hearing. After that, the builder should have maximum latitude within the constraints of physical environmental limitations and public facilities to proceed.
4. We need to explore and understand one another's problems. I imagine builders would better understand environmentalists if more of the former served on school boards, took part in open space programs, and fought for environmental quality as hard as they fight for their own projects. Of course, on the other hand, environmental groups such as the Sierra Club would have a better understanding of the cold economic realities that builders face every day if they undertook to construct 1000 low-income dwelling units per year.

The Road from Calvert Cliffs to Mammoth

M. R. BEYCHOK

Since 1970, environmental legislation, like emancipated womanhood, has come a long way, baby! If you plan a major new industrial plant in the United States today, you must travel through a strange new land, full of court decisions with names like *Calvert Cliffs*, *Greene County*, and *Friends of Mammoth*; you will encounter the Louisiana offshore lease sales, the Alyeska pipeline, the nondegradation clause, and other anomalies. You must also learn a new language with words like NEPA, EIS, primary and secondary ambient air standards, dispersion models, implementation plans, OSHA, and emission standards. And above all, you must learn to read an unusual newspaper called the *Federal Register* as thoroughly as you once read Perry's *Chemical Engineers' Handbook*.

This chapter presents a brief guided tour through some of this strange new land to help foster an understanding of the current environmental legislative scene. I will try not to pass judgments or to exhibit my personal bias; my only intent is to make the reader aware of the many complexities and obstacles resulting from recent legislation.

The Clean Air Act and its Implementation

When Congress enacted Public Law 91-604, the Clean Air Act of 1970, the law instructed the Environmental Protection Administration (among other things) to establish national air quality standards. The law also gave EPA the legal authority to see that each state developed Implementation Plans for attaining and/or maintaining the national air quality standards. If the states

Mr. Beychok is the Manager of Environmental Engineering, Fluor Engineers & Constructors, Inc., in Los Angeles.

failed to develop adequate plans or failed to enact the appropriate legislation to enforce their plans, EPA was authorized to step in to revise and/or enforce appropriate Implementation Plans.

At this point, it is most important to fully understand the difference between ambient "air quality standards" and point source "emission standards." Air quality standards set the level of purity desired for the ambient atmosphere as a whole. In other words, the ambient atmosphere is to contain no more than 100 micrograms of NO_x per cubic meter of air (as an annual mean). Emission standards, on the other hand, limit the amount of contaminants to be discharged from a point source plant or vent stack, usually measured at the point of emission and expressed in pounds per unit of plant input or output. The setting of emission standards must be low enough to ensure that the ambient air quality standards will be attained or maintained. For example, if we set NO_x emission standards for all stationary fuel-burning equipment stacks at, say, no more than 0.3 lb of NO_x per MM Btu of heat release, and if we set NO_x emission standards for automobiles at no more than 3.0 g per vehicle mile, perhaps we can attain or maintain the NO_x in the ambient atmosphere at no more than 100 $\mu g/m^3$. Emission standards are the point source control "tools" by which we attain or maintain the desired air quality standards.

By April 1971 EPA had promulgated national air quality standards for six air contaminants: sulfur dioxide, NO_x, particulates, carbon monoxide, and photochemical oxidants. These were set on the basis of scientific and medical studies and after fairly broad-based public hearings. Some of the standards were attacked by some segments of industry as being too stringent. But in general, I believe it can be said that *no one really understood the full economic and technical implications* of the air quality standards when they were set in April 1971. As directed by Congress, EPA set the standards to protect human health, and the agency probably did the best it could with the very limited scientific and medical data available, and in the limited time allowed. Very few states at that time knew what their air quality was, thus did not know how the standards would affect them. In fact, many states had no effective, adequately staffed and funded agencies devoted to air quality or emission control. In general, prior to 1971 industry took little active interest in assessing the impact of the national air quality standards on future industrial growth.

Having set the air quality standards, EPA instructed the states to prepare their Implementation Plans for attaining or maintaining the standards. The required elements of the Implementation Plans were set forth on August 14, 1971, in the *Federal Register*. Among other things, they included the following requirements:

- Each state was to determine its existing air quality. If measured data were unavailable, the states undertook a point source emission inventory and then "estimated" their existing ambient air quality using the emission inventories and air dispersion calculational models.

- If the existing air quality was poorer than the level set by the national standards, a state was to establish a compliance schedule indicating it intended to attain the national standards.
- Each state was to set forth in the Implementation Plan the specific control measures, emission standards, and so on, by which it intended to meet its compliance schedule, or maintain its clean air if the existing air quality was better than the national standards.
- Each state was to enact the legal authority for new emission source construction permits (if it did not already have such legal authority). *AND THOSE NEW SOURCE CONSTRUCTION PERMITS WERE NOT TO BE ISSUED IF THE NEW SOURCE INTERFERED WITH ATTAINING OR MAINTAINING THE NATIONAL AIR QUALITY STANDARDS.*

I have capitalized and italicized the last item; it is unfortunate that the *Federal Register* did not do the same. In fact, it is unfortunate that our technical and trade journals did not front page and editorialize that item. Not only did we all fail to see the potential impact of that permit criterion on August 14, 1971, but all too few people are aware of its implications today!

By early 1972 almost all the states had sent their Implementation Plans to EPA. After some months of review, EPA issued its revisions in a series of promulgations in the *Federal Register* during the period from May through October of 1972 (the exact dates and *Federal Register* references for all these events are chronicled at the end of the Chapter). And *now* it will become clear why I keep using the words "attain or maintain the national air quality standards": after the EPA revisions of the Implementation Plans, 30 to 35 states had some airshed or air basin regions that were currently violating the national air standards. That is, their air quality was poorer than the national standards permitted. Thus 30 to 35 states were faced with "attaining" the national air standards in some parts of their respective jurisdictions. Unsurprisingly, the 30 to 35 states include most of our industrialized areas.

Suddenly it became apparent to some state and local regulatory agencies—New Mexico and California, for example—that *legally* they could not grant any new plant construction permits if the EPA promulgations were allowed to stand. If the criteria for granting new construction permits included noninterference with attaining national air standards, and if the state were already above the national standards, the new plant would interfere with attaining the national standards even if it emitted one-tenth of the contaminants emitted by similar existing plants of similar size, or even if it utilized all known technology and met or bettered all existing emission standards. In fact, no matter how low the new plant emissions were, they would "interfere with attaining the national standards."

To the best of my knowledge, only a handful of states have strongly protested the situation as it has now become apparent. In the Four Corners states (New Mexico, Arizona, Utah, Nevada), some half-dozen industrial companies have filed court actions to contest the situation with EPA. In California, the Los Angeles Air Pollution Control District (APCD) has protested strongly[1] against "a no-growth policy,"which is said to be the result of the "unwarranted, low ambient air quality standards." As we can now see, no one fully understood the resultant economic implications of the air quality standards when the EPA set them in April 1971.

In summary, this is where the Clean Air Act has brought us:

1. We have national air quality standards.
2. In early 1973, there were 30 to 35 states in violation of those standards (in part or all of their air basins).
3. We have a permit system that prohibits new plant construction if emissions from the new plants interfere with attaining the national air quality standards.

There are those who speculate that EPA or some individual states will not fully enforce the letter of the law in all cases, or will use due discretion. However, we can be sure that the public interest environmentalist groups (Sierra Club, Environmental Defense Fund, National Resources Defense Council, etc.) will exert pressure and use litigation to enforce the letter of the law.

What can be done? That is up to you—first to pass judgment on the situation, then to take appropriate action. My position is simply that too few people are aware of the legal situation. I want more people to know what has happened.

What Is an EIS?

It might be simpler to start by telling what an Environmental Impact Statement EIS is *not*. It is *not* a 10 to 15 page report; it is *not* a 1 to 2 week effort; and it is *not* a low-cost item. It is a major undertaking that may require 6 to 18 months for preparation, circulation among governmental agencies, public hearings, and approvals, and for a major new plant it may cost as much as $100,000–$200,000.

Having stated what an EIS is and is not in terms of time, effort, and cost we now ask, what is an EIS? It is an Environmental Impact Statement (EIS) or Environmental Impact Report (EIR) or Environmental Impact Assessment (EIA).

The National Environmental Policy Act of 1969 (NEPA) required the full public disclosure of the environmental impact of any action taken by a federal agency. Specifically, Title 1, Section 102, of NEPA requires that "major Federal actions significantly affecting the quality of the human environment" be preceded by "a detailed statement by the responsible official" which shall include:

1. The environmental impact of the proposed action.
2. The adverse environmental effects that cannot be avoided.
3. The alternatives to the proposed action.
4. The relationship between the short-term and long-term effects on the environment.
5. Any irreversible or irretrievable commitment of resources.

By early 1973, about 2900 Impact Statements had been developed regarding actions by federal agencies, and these resulted in more than 200 lawsuits.[2]. This is where we encounter some of those strange places and names, which have determined some of the key court interpretations regarding EIS requirements. The citizens of Calvert Cliffs, Maryland, won the first landmark court decision, against the Atomic Energy Commission. The court ruled that the AEC had no right to proceed with nuclear plant projects before preparing an EIS and could not neglect water quality or other impacts simply because other state and federal agencies had already certified compliance with their effluent and emission standards.

The Planning Board of Greene County, New York, won another key decision, against the Federal Power Commission. The court ruled that the FPC must prepare its own Impact Statement prior to approving a high-voltage transmission line and could not simply use the project applicant's statement.

When the Department of the Interior proposed to sell oil and gas leases in offshore Louisiana, the Natural Resources Defense Council won a court injunction because the department's EIS had an inadequate discussion of alternative actions.

In another court case, the Corps of Engineers was enjoined from issuing waste water discharge permits unless an EIS was prepared prior to granting permits.[3] Another court decision, *Sierra Club* vs *Sargent,* affirmed the same point.

The proposed Alyeska crude oil pipeline across Alaska is currently in court, being challenged by a number of environmentalist groups claiming, among other things, that the Department of the Interior's EIS gave inadequate consideration to alternate routes.

California has a law quite similar to NEPA, the California Environmental Quality Act of 1970 (CEQA), which also requires Impact Statements as part of the decision-making process before project approvals and building permits are granted. In a landmark suit brought by the Friends of Mammoth against the County of Mono, the California Supreme Court ruled six to one that even private housing projects required an EIS. This recent decision (September 21, 1972) has created much confusion and many building permit moratoriums all across California. Another suit under CEQA, *Environmental Defense Fund* vs *Coastside County,* has been decided in favor of the plaintiffs. The court ruled that an EIS must consider all phases of an integrated project and can not avoid evaluating the overall impact, including future phases of a staged project.

All this litigation has led to certain key interpretations regarding the preparation and content of an EIS. Let me list some of them:

- The EIS must be developed by the lead governmental agency involved in approving a project, not by the project sponsors or applicants. The applicant may be and probably will be called on to furnish baseline preconstruction environmental surveys, to describe the project technology and the resulting emissions and effluents, and to evaluate alternatives. However, the final EIS must be written by the agency, which must use its own value judgments.
- The most important single aspect of an EIS is the consideration given to alternatives. The adequacy of alternatives examined is the point most susceptible to court litigation.
- No agency can exempt itself or exempt any specific project from requiring an EIS, without being subject to court litigation. The laws, both NEPA and CEQA, warn that an EIS is required only for "major, significant" actions. In practice, litigation seems to have forced an EIS wherever citizen opposition has gone to court. Even the granting of waste discharge permits or private housing building permits have been ruled to require an EIS.
- The EIS must be a broad interdisciplinary study of the total integrated project. The environmental impacts to be considered are not only those associated with air, water, and noise; they must include sociocultural effects, economic impacts, transportation effects, archaeological and paleontologic impacts, and aesthetic impacts. In other words, an EIS must consider all impacts on the human environment as well as on the physical environment. The preparation of an adequate EIS for a major new project will involve many disciplines other than that of the engineer. Marine biologists, life-science biologists, geologists and hydrologists, economists, social scientists, urban planners, and archaeologists are all necessary.

An EIS is not a brief report and will require far more than a few weeks to prepare. It is a major undertaking and one with which we had all better become intimately acquainted.

The Nondegradation Clause

The Sierra Club recently won a case[4] against EPA (Sierra Club vs Ruckelshaus) which has very broad implications related to what I have discussed thus far. As you will recall, the EPA promulgations under the Clean Air Act have established a legal basis for not granting construction permits that would interfere with "attaining" the national air quality standards. And we must keep in mind that some air basins in the nation still have an air quality that is *better* than the national air standards. The Sierra Club sought to establish that the Clean Air Act did not give EPA the right to sanction degradation of clean air *down to the national air standards* . . . where existing air quality exceeded the standards. The court decision affirmed this principle.

We now find that we cannot build a new plant that will interfere with "cleaning up a dirty air region," nor can we build a new plant that will "degrade a clean air region." In effect, new industrial growth is caught between the proverbial rock and hard place.

The New Federal Water Pollution Control Act[5]

On October 17, 1972, the Congress overrode a presidential veto and wiped the slate clean of many previous water bills by enacting the new Water Pollution Control Act of 1972. This very comprehensive act includes a host of major new programs. One of its stated policy objectives is to completely eliminate discharge of pollutants into the nation's waterways by 1985. To achieve this goal, effluent discharge limitations based on "best practicable technology" will be established by July 1977, and effluent discharge limitations based on "best available technology" will be established by July 1983.

We are not yet ready to fully interpret or understand all the provisions of the new act. But here is a partial summary of its contents:

Title I. Deals with research, training, and information programs and provides for their funding.

Title II. Provides funding for EPA to make construction grants for municipal and industrial waste treatment plants.

Title III. *Standards and Enforcement*

Section 301. EPA to establish 1977 "best practicable technology" point source effluent standards, and 1983 "best available technology" point source effluent standards.

Section 303. EPA to review the states' receiving water quality standards and Implementation Plans for attaining those standards (in effect, gives EPA similar powers as for national air standards).

Section 304. Within 1 year, EPA is to develop and publish:

1. Criteria for receiving water quality standards, reflecting latest scientific knowledge.
2. Factors necessary to restore the integrity of the nation's waters.
3. *Effluent Guidelines* for use in adopting the 1977 and 1983 *Effluent Standards* in Section 301.

Section 306. EPA to identify and publish a list of effluent sources from new plants in 27 different industries.

Within 1 year, EPA to establish *new plant* Performance Standards based on "best available technology" for the new plant sources in the 27 different industries. (These Performance Standards for new plants *may* include zero pollutant discharge.)

Section 315. Establishes a National Study Commission to report on the technical, economic, and social effects of achieving the 1983 goals set in Section 301.

Title IV. *Certificates and Permits*

Section 401. Establishes new plant construction certificate program, certifying compliance with the act. If states do not already have such certificate program, EPA to establish its own.

Section 402. Establishes a waste discharge permit program, administered directly by EPA.
Regarding the 1899 Refuse Act permit program, this act replaces the 1899 act.

Title V. *Administrative*

Section 503. Establishes a Water Pollution Control Advisory Board to advise EPA on all functions under the Act.

Section 515. Establishes an Effluent Standards and Water Quality Information Advisory Board to advise EPA on proposed regulations under Sections 304 and 306.

Section 511. Exempts EPA from writing Environmental Impact Statements regarding its own actions under the act, except for:

1. Construction of waste treatment plants under Title II grants.
2. Waste Discharge Permits granted under Section 402 for *new* sources defined in Section 306. . . (presumably so as not to contradict the *Kalur* vs *Resor* decision).

If anyone had doubts before, he surely realizes now that we face essentially zero discharge of pollutants as a requirement for existing plants by 1983 and perhaps even sooner for *new plants*, since the new plants must meet the "best available technology" criteria rather than the "best practicable" criteria.

California Coastal Water Discharge Quality

On July 6, 1972, the California State Water Resources Control Board published a Water Quality Control Plan. Tables A and B of Chapter IV set forth the effluent quality standards for waste discharges into coastal waters. They are quite stringent. In fact, some knowledgeable people in the field of waste water treatment have said that tertiary treatment will be required to achieve these standards. I myself question whether a total chromium limit of 0.005 ppm, for example, is even within the realm of possibility—this value is 10 times more stringent than the U.S. Public Health Service standards for drinking water.

Since very few municipal waste treatment plants in California have secondary treatment, much less tertiary treatment, it has been estimated that compliance with these discharge standards may cost California municipalities several billion dollars.

Documentation of Events

As authorized under the Clean Air Act of 1970 (Public Law 91-604), EPA has done the following:

1.0 Promulgated national ambient air standards (*Federal Register*, April 30, 1971), which include a number of categories.

2.0 Promulgated requirements that each state submit Implementation Plans setting forth how the state would attain the national ambient air standards (*Federal Register*, August 14, 1971).

Section 420.11(a)(4) specifically requires each state to acquire legal authority to "prevent construction of any stationary source where emissions from that source would prevent attainment of national air standards."

3.0 Received and reviewed the Implementation Plans from each state, advised which parts of the plans were disapproved, and promulgated EPA's revisions to those plans. (*Federal Register,* May 31, 1972, July 27, 1972, September 22, 1972, all entitled "Title 40—Protection of Environment.")

In most cases, the disapproved Implementation Plans included tabulations of which regions in the states were *exceeding the national ambient air standards*.

Timetables were set for those regions to take corrective action so as to *attain* the ambient air standards. For example (and this is not a complete listing), the states having regions now exceeding the national air standards in one or more categories include:

New Mexico (e.g., Four Corners region)
California (e.g., Los Angeles)
Texas (Corpus Christi, Houston, and areas)
Louisiana (entire state)
Pennsylvania (almost the entire state)

4.0 Promulgated and adopted regulations *to be enforced by EPA* that would ensure that the revised plans would *attain* the national air standards (*Federal Register*, June 14, 1972, July 27, 1972, September 22, 1972 all entitled "Environmental Protection Agency, 40 CFR Part 52").

In many cases, these regulations include:

(a) The construction and design of new sources (i.e., new plants) require a *permit from EPA*.
(b) Permits *will not be issued* if the new source "interferes with the *attainment* or maintenance of any national air standard."

Thus here are three examples:

New Mexico

(a) *Federal Register*, July 27, 1972, page 15087, Section 52.1630 (table therein) establishes that the Four Corners area is now exceeding air standards.
(b) *Federal Register*, July 27, 1972, page 15107, Section 52.1628 states that new sources need EPA permits and permits will not be granted if new source interferes with attaining national air standards.

California

(a) *Federal Register*, September 22, 1972, page 19813, Section 52.238 (table therein) and May 31, 1972, page 10854, Section 52.238 (table therein) establishes regions now exceeding air standards.
(b) *Federal Register*, September 22, 1972, page 19833, Section 52.233 states that new sources need EPA permits and permits will not be granted if new source interferes with attaining national air standards.

Louisiana

(a) *Federal Register*, May 31, 1972, page 10870, Section 52.979 (table therein) establishes regions now exceeding air standards.
(b) *Federal Register*, June 14, 1972, page 11832, Section 52.976 states that new sources need EPA permits and permits will not be granted if new source interferes with attaining national air standards.

5.0 In some states (e.g., Texas and Pennsylvania), EPA has promulgated that the national air standards are being exceeded (see 3.0), *but* EPA has *not* promulgated an EPA permit system within those states. I *assume* that those states already have their own new source permit systems which prohibit permits that will interfere with attaining national air standards (but I have not researched this point). If my assumption is correct, the situation vis-à-vis new source permits is legally the same in those states as in the ones where an EPA permit system was established.

Notes And References

1. Letter from APCD to EPA Regional Office, October 18, 1972.
2. Third Annual Report by the Council on Environmental Quality (to the President, as required under NEPA).
3. *Kalur* vs *Resor*, U.S. District Court of Appeals, December 1971.
4. *Sierra Club* vs *Ruckelshaus*, U.S. District Court, District of Columbia, June 2, 1972.
5. Completely replaces (1) 1965 Water Quality Act, (2) 1966 Clean Water Restoration Act, (3) 1970 Water Quality Improvement Act, (4) 1956 federal Water Pollution Control Act.

Marketing Environmental Management

R. PAGE JONES

To assist people, both citizens and decision makers, to make sound and well-informed decisions on land use and pollution control, a substantial and effective communications component must be employed and carefully integrated. The people must first be aware of the problem to which the decisions relate; the problem must be relevant to their personal aspirations and concerns; they must perceive one or more courses of action that seem practical and worthwhile; and they must be willing to act. The essential steps are: develop awareness, create interest, build support for solutions, and get action.

At the start I must insert two disclaimers. First, I base these remarks on the knowledge and experience of my organization. We have wide experience, yet we would be the first to admit that informing, persuading, and motivating the public is far from a science and that, indeed, for every solution there is a problem.

My second qualification—or caution—is that communication cannot help, and may actually be harmful, unless those who seek action have clearly defined the problem, know exactly what they want to accomplish, and know why.

Awareness and Interest

Now to the first step, developing awareness. This has already been done. In 1972 a national opinion survey, carried out in great depth by the Gallup organization, indicated that Americans are very much aware of air and water pollution and of the problems of solid waste disposal. In this study, environ-

Mr. Jones is President of Phillips-Ramsey Inc., San Diego's largest advertising agency.

mental problems ranked near the top of public concerns. By comparing this poll with a study conducted in 1968, we learn that public concern about the environment has increased 23 percent in just four years—an extremely sharp upward tilt. Those *most* concerned included the young rather than the old, the well educated rather than the poorly educated, the more affluent rather than the poor, professional and business people rather than those who are retired or unemployed, westerners rather than residents of other sections of the country, city dwellers rather than rural residents, whites rather than blacks, Catholics rather than Protestants, independents rather than persons who identify with the two major political parties, and liberals rather than conservatives.

Significantly, along with great awareness, there is great interest. Thus our problem, the problem of communication, boils down to enlisting support for proposed solutions, and more important, action.

In my opinion this interest and awareness has in itself created a very difficult situation, because I also feel that public expectations are seriously inflated. As Americans, we are all too aware. We want something done about the problem. We do not know just what. We would prefer that it cause no great inconvenience, such as restricting our freewheeling mobility. And we certainly hope that it can be achieved at no great cost, such as would call for an appreciable increase in taxes.

Building Public Support

Our challenge is not simply to communicate but to communicate reasonable solutions. We must educate the public on just what is *rationally* possible in the early universal desire to cope with pollution.

Somewhere between the environmental extremist and the stereotyped industrialist, leaders must find the middle ground and have the courage to plan rational growth policies, to stand up for them, and to embark on the difficult and tedious course of convincing their fellow citizens of the necessity and wisdom of their proposals—realizing that no one will be entirely happy with the results.

Right now the public desperately wants to see environmental problems solved, but people are confused almost beyond belief by the complexities, the counterclaims, the dire warnings, the almost hourly statements of explanation that clarify nothing at all.

I believe that a vital first step to effective communication must be some solid thought and some necessary compromises; the creation of a course of action that can be communicated with confidence.

There is a crucial distinction between selling and marketing. In a sales program, we simply start with a product and try to persuade people to buy it.

In efforts to market environmental programs, one must start with what the public wants and needs—a program that is reasonable, understandable, believable, and consistent.

That is a terribly difficult challenge for three important reasons. It is hard to communicate anything today because (1) we are already overloaded with information, (2) it is not easy to tie a message to the interests and concerns of an audience, and (3), the changing of deep and tightly held beliefs is often impossible.

Today we are so heavily bombarded by words that we set up unconscious defenses. We simply do not see or hear much that is aimed at us. We filter out what we find irrelevant. As a result, it is getting more and more difficult to use mass media effectively. To cut through the clutter, it is increasingly important to appeal to the self-interest of our market. If you want to persuade people to support integrated environmental management, for example, you must tell them in terms as specific as *possible*. All too often communicators feel that everyone is automatically interested in their problems.

A large percentage of news releases, statements, booklets, and other materials issued by government and business suffer from two deadly flaws: they are written from the communicator's not the reader's, point of view, and they assume a knowledge of technical or bureaucratic jargon that the reader does not have.

Compounding it all is the fact that people are very difficult to change. Most of us assume that if people are properly educated, propagandized, and informed, they can be induced to change their habits or beliefs. Generally this is not true. Persuasive words change very few habits, usually only those that are lightly held. Consider attempts to alter smoking, drinking, and driving behaviors. In short, the strength of deeply held habits and beliefs must be taken into account in planning effective communications.

A Communication Action Plan

In light of this brief background, here are 10 specific suggestions.

1. *Have definite audiences in mind.* You do not simply communicate with people. You talk to certain groups.
2. *Develop a basic theme or statement.* Be consistent. Summarize your story in a page or less. Once that statement has been refined, use it over and over. Use the same words and the same arguments until you are thoroughly sick of them. Because by then your audience may have begun to get its first faint glimmer of your message.
3. *Decide on a budget.* Effective marketing campaigns are expensive. If

possible, use well-qualified people to plan and carry out your program. It may save you money in the long run, because a clumsy or ineptly handled program is often counterproductive.

4. *Decide on a time frame.* If you must accomplish something, set a date for accomplishment. Ideally, you should have several dates in your plan, with definite objectives to be met at each stage.

5. *Use as much research as possible.* To undertake a marketing program without research is like trying to drive from San Diego to Montreal without a map—it is possible but foolhardy. Research has two important functions. One is to help set your course, to tell you what your audiences think and how they feel about the issues. The other is to help evaluate what you have done, showing which activities have been worth the time and money, and which have not.

6. *Try to create big news.* People are literally flooded today with news and information. They tend to overlook anything that is not stated in strong terms, even sensational terms. Conversely, most government officials tend to issue statements that are carefully qualified, understated, and calculated to upset or offend no one. Unfortunately, the nice quiet news release dies a nice, quiet death.

7. *Concentrate on people-to-people activities.* I think this is the most important advice I can offer. The most effective way to communicate is person-to-person, not through printed or electronic media. Then why do the big companies spend fortunes in mass media? They have no choice. But the environmentalist quite probably does have a choice.

You can work hard to arrange meetings of interested persons by appealing to their self-interest. You can work through existing organizations. And you can help create new groups that have a special concern in spreading your message. You can establish two-way communication, which is the very best kind, because it allows people to feel that they are participating, that your cause is theirs.

Admittedly, it is much more difficult to mount a marketing or public information campaign through the use of person-to-person strategies than it is to use mass media. There is an understandable feeling of accomplishment when you receive a handsome brochure from the printer, even though no one beyond your personal staff pays much attention to it. There is similarly a wonderful euphoria when you issue a news release or send out a television spot. All too often, however, we begin to regard the message itself as the goal, rather than the change in attitudes and behavior that we are aiming for.

If you really want accomplishment, not just the illusion, try to execute as much of your program as possible through intimate, forceful, person-to-person encounters.

8. *Use volunteers.* Despite the increasing talk about voluntarism, a relatively poor job has been done in utilizing the great reservoir of volunteer power available in the United States. In most major civic or community undertakings, the same people are used over and over, while vast numbers of highly motivated and highly qualified people are not used at all. Why? Mainly because most of us do not take the trouble to look for a fresh face. A San Diego councilman recently selected the members of a citizen advisory board by picking them at random from the 8500 persons living in the neighborhood. They were asked to serve three-year terms. Of those selected, 90 percent responded. Most of those selected had wanted some civic involvement but had never been asked.

9. *Work hard to unite community leadership.* Communication is a two-step process. Usually the source of the information supplies the message to opinion leaders or persons of influence, and these people in turn inform their followers. Influential individuals are not necessarily the big names. Very often they are long-time residents of a neighborhood, factory foremen, or bartenders in local hangouts. But to the extent that you can identify these leaders and reach them, your task will become easier.

10. *Use mass media effectively.* Mass media are not worthless. I have tried to offset the overreliance that is generally placed on communication by newspaper, radio, television, and billboards. If you want to communicate, mass media, provided they are professionally utilized, can be a very effective *supplement* to the face-to-face techniques.

I hope that these suggestions can assist you in doing a better job of marketing environmental management to the general public. None of the proposals is complex. Collectively, they offer no magic formula. None provides an easy answer or solution.

Communication is hard work.

Part VIII

Environmental Education

Thomas A. MacCalla

Dr. MacCalla is Director of the Center for Human and Urban Development at United States International University, San Diego.

Usually the term "environmental education" is associated with public debate on environmental policies and practices, with curriculum innovations in schools and colleges, with Earth Day, or with other public awareness programs. However, we should be conscious that environmental education is all these things and more. Because it is all-inclusive, we need to define the term broadly enough to express our intent and specifically enough to provide meaning. We want to consider the size, shape, and style of "environmental education."

As we know, the word "education" refers to the actualization process or the drawing out of human potential. Education, though, is also a man-made form or combination of learning structures designed to perpetuate value systems and to impart new knowledge, skills, and attitudes to the society's members. Similarly, we define the word "environment" to mean all the surrounding conditions and influences that affect the development of living things. In short, environment is both natural and man-made, and environmental education is learning to function within that context.

Using the term broadly, let us consider environmental education as *the process of creating individual and collective commitments to improve the quality of life through self-knowledge and an understanding of the physical, political, socioeconomic, and behavioral constraints on living.* As such, we are able to explore freely the many dimensions inherent in our topic. For our purposes we have sampled some of the ideas and forms that provide useful meaning to the term. These final chapters are organized around the strategy of information giving, information receiving, and information exchange.

Our topic can be defined as schooling and community participation, as community communications or ways of reaching the various publics that constitute the community, as the framing of a new perspective on diagnosing and managing the sociopsychological and physical variables in the human environment to ensure man's health and well-being, as the holistic view of community, collaborative planning, and institutional responsibility for training change agents and practitioners, even as regional and binational environmental health programs that transcend geographic and cultural boundaries because problems and issues are interrelated and people are interdependent.

I want to emphasize that environmental education focuses on more than the physical concerns of man. Moreover, environmental education has a role as a continuous, individual, and community education process that is an integral part of the complicated war of communications for human understanding. May I even suggest that unless we recognize and comprehend the many facets of the medium and the message, we will not be able to create the socially supportive environment needed to wage the battles successfully.

Man shapes his environment and is shaped by it, and the process involves the realities of living in an urbanized and culturally pluralistic world community. Hence we must recognize that internationally, the prescriptions of the highly industrialized nations to achieve environmental quality could be perceived as being defensive, monopolistic, or even oppressive measures for managing the future of the less-developed or emerging nations in the world.

Nationally, we must contend with conflicting professional advice, legal decisions, and public petitions on regional growth, progress, and human development. We must, therefore, acknowledge not only the position that planners plan with people and not for them, but we must also be more sensitive to the political, economic, and sociocultural realities and consequences of environmental decision making.

To validate the administration's call for widespread public concern about environmental issues or to carry out the pronouncement asking for new knowledge, new perceptions, new attitudes and a basic reform in the way the nation looks at problems and makes decisions, we must contend with the contradiction of reduced financial support for programs needed to honor that commitment. We cannot stress the need for resolving the energy and fuel crisis or arresting pollution problems without placing equal emphasis on regional and binational approaches to community environmental management or environmental health education and related human care services.

We must be responsive as well as responsible. Our emphasis on regional policies, cost analysis, or economic impact of environmental management, environmental quality indices, and information systems must be considered to be part of the whole. People, the human element, must be central to all our deliberations. We need to move from rhetoric to program, providing the necessary time, talent, and money. New categorical aid or support through revenue sharing notwithstanding, we need to forge consortia of public and private interests such as are represented at this conference to provide the content for environmental education.

A Community-Centered Program

JOHN K. GESSEL

This chapter explains the process of involving community members in the development of an environmental education program aimed at teachers of students in the first through twelfth grades.

Rationale for the Model

There are two reasons for involving many community members in the development of a community-centered environmental education program model. The first is that many local people have a need to educate the adults on a short-term basis because of immediate community decisions that have to be made. By involving adults in the preparation of a kindergarten through twelfth grade curriculum, this becomes a mutually reinforcing system. The adults bring forth their expertise, work with other community members using the Center for Environmental Education as an operational vehicle, and the center gains information from the community members for environmental education curriculum development.

The second reason is that a long-term educational effort has to be mounted for teaching students about environmental problems. Therefore, the teachers of school-age students become the priority target population.

The concept of this model is broad, because as more people become involved, more target populations are reached. The ultimate target populations the

Mr. Gessel is the Director of the Office of Community Education Resources in the San Diego County Department of Education. He also managed the HEW-funded Center for Environmental Education Project.

model intends to reach include the community at large, decision makers within the community, service organizations (in the broad sense), and educational organizations.

Process for Implementing the Model

The model was operational for one year in a county in California where about 15,000 teachers were working with students in grades 1 through 12. One of the 48 school districts in the county, the large city school district, accounted for about one-half of the teachers. The other 47 districts ranged in size from less than 10 to more than 1000 teachers in each of the various districts. The geographical distribution of the schools throughout the county is most dense along the western boundary, becoming increasingly sparse toward the eastern boundary. The county contains 4791 square miles. The total population of the county exceeds 1.3 million.

To engage a community of this size in a meaningful way required a great deal of preplanning, long-range planning, and short-range planning. The preplanning took the form of an ad hoc group, representing a broad range of the community. There were no more than 12 members on such an ad hoc committee, who were individuals from the middle-management levels of business, the military, higher education, and the schools. This group formulated conceptual notions of how an environmental education program should evolve. The time between each meeting allowed the members a chance to talk with others, return with additional information, and begin to develop a commitment to the concept of environmental education.

A further purpose of the ad hoc group was to identify other organizations, and people within the organizations, who would eventually be brought in as a part of the evolution process. More than 50 community organizations eventually became involved with the Center for Environmental Education. After three or four ad hoc meetings over a span of as many months, a larger group meeting, attended by about 50 people, became a necessity. The theme for the large group meeting focused around two areas: what curriculum content should be included in an environmental education program, and who should be responsible for each major area of development? The meeting was a two-day weekend retreat; the participants were totally committed to discussing the topics and interpersonal relationships had a chance to form. During this meeting recorders were assigned to listen to the small group sessions and report their summaries during the large group sessions. The resulting document became a framework for a community-centered environmental education program.

Once a framework was available, it was appropriate to enter the long-range planning phase. This was accomplished by calling a meeting for everyone interested in environmental education. Invitations were widely broadcast to every agency in the region that might be interested in a program of environmental education. At this meeting the framework was presented by asking how each agency might help with the implementation of the plan.

To answer the question, the large group was divided into small discussion groups for an exchange of ideas. Forms were available for agencies to fill out indicating how they could help. When the small groups reassembled as a large group, they summarized the discussion of the small groups. Also, ample time was set aside for questions. This meeting was chaired by members of the original ad hoc committee.

At this meeting, nominations were made from the large group to create a steering committee that worked with the educational staff. The members were again representative of various community sectors, but the committee did not have more than seven categories. The steering committee and education staff met as an operational group to consider the information provided by the agencies for implementing the framework. The decisions made at these meetings constituted long-range planning.

Short-range planning began to emerge as projects were identified that could be implemented within a year's time. It was most important to be able to define precisely what was to be done, so that a report could be made to the large group of citizens regarding the decisions of the steering committee. When these decisions had been made, a clear picture was presented to the community members, who could see how to become involved.

Through their involvement in project implementation, community members became more highly educated in environmental issues at the same time they were teaching others. An example of this model is the Center for Environmental Education (CEE), which is depicted in Figure 1.

The large community group just discussed was represented by the CEE Advisory Council; any person in the community could become a member simply by attending a meeting. The elected steering committee was called the Executive Board of the CEE Advisory Council for this model. The CEE Executive Board worked closely with the education staff. The CEE Operational Steering Committee was invited to meet once or twice a year to review projects relative to how they fit into an educational perspective before they were passed on to the Superintendent of Schools for final approval.

Other umbrella organizations that represented human resources are listed as individual citizens, the Integrated Regional Environmental Management (IREM) project and Industry-Education Council, and the Community Educational Resources project.

The model listed the 31 projects in which the community became involved. The Appendix details the projects, which all lent themselves to environmental education. Finally, the target populations for education in the area of environ-

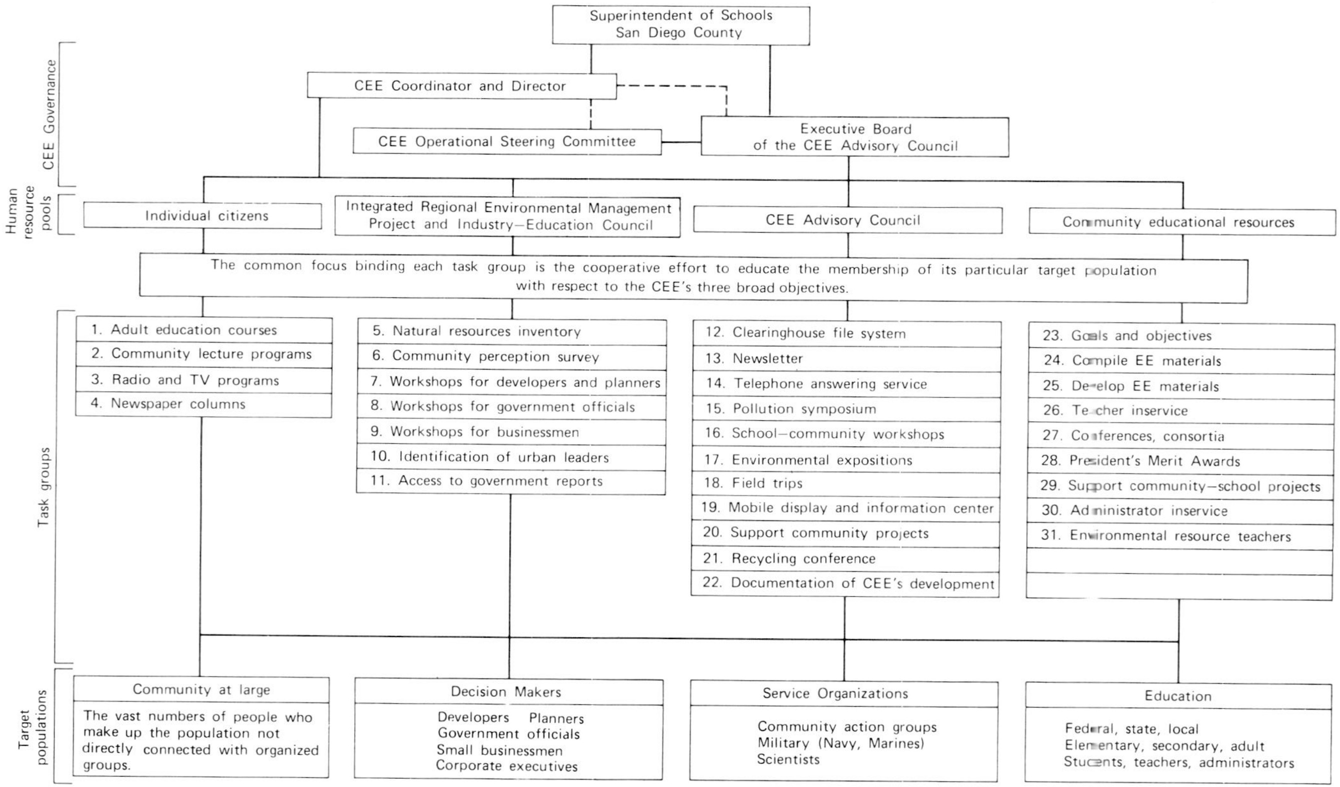

Figure 1. Center for Environmental Education model.

mental issues were enumerated at the bottom of the chart to include the community at large, decision makers, service organizations, and educators at all levels.

This operational scheme came to represent the long- and short-range plans for an environmental education program. Not all the projects could be started or completed within a year, but projects were added and deleted with the passage of time.

Evaluation

The projects that were started during the first year of operation of the CEE were evaluated using questionnaires. Generally, the projects deemed most important by both community and educators were: teacher inservice, curriculum development, the newsletter, the quick access file, and the speakers' bureau.

The pros and cons relating to each of the tasks are as follows:

College Credit Courses. All classes were filled to capacity, but expenses were very high, and without outside aid, the courses could not be continued. (Task Group 26)

Other Teacher Inservice. The Coca-Cola recycling conference was so successful that it was scheduled for repetition. Action programs started by teachers as a result of this conference have been documented. The "Education for Environmental Awareness" conference was successful, but will not be conducted again because so much volunteer help was enlisted. It took one person to spark the event and without such leadership, the event cannot be repeated. "Cycles of Life in the Sea" will probably be run for a third year because the costs are minimal, interest is high, and leadership continues. (Task Group 26)

Community/Schools Workshops. This was one of the most successful programs but required a great deal of staff coordination. Without a paid coordinator, this kind of function was impossible. The costs to participants were low and could have been met by them, but with no coordinator the program would not run for a second year. (Task Group 16)

Environmental Fairs. These events required a great deal of time, and their value was questionable in proportion to the time spent. One fair in a metropolitan shopping center could be justified. (Task Group 17)

Newsletter: "Environmental News and Views." The success of the newsletter was measured by the increased demand for the publication from 20, 000 to 28,000 copies over five issues. The community felt the newsletter to be so important that it was continued and published by the IREM project of San Diego County.* (Task Group 13)

Field Trips. This function was important but not of high priority. An important result was a publication listing field trips for teachers' information. The material will be further developed and given wide distribution. (Task Group 18)

Visiting Lecturers. This project was not worth the time and money needed for its implementation. The main purpose was to educate the general public, and not enough people attended the sessions. The mass media forms of communication must be used to achieve the purpose of educating the general public. (Task Group 2)

Compilation of Environmental Education Material. An essential component of any project for referencing and maintaining a current knowledge in the field. (Task Group 24)

Other Projects

Speakers' Bureau—used continually by teachers. (Task Group 14)
Guide to Community Resources—used in conjunction with Speakers' Bureau. (Task Group 14)
Review of Local Audiovisual Materials—limited use until wide distribution was made. (Task Groups 3 and 24)
Evaluation of Printed Materials—very useful to staff when recommending materials. (Task Group 24)
Quick Access File—limited use until wide distribution was made. (Task Group 12)
Lagoons Paper—was discussed in *Environmental News and Views* and was added to the quick access file. (Task Group 24)
Spanish Translation of Environmental Brochure—translation was made with the understanding that the State Department of Education would publish it; not yet published. (Task Groups 10 and 25)
Screening TV Programs—programs were selected and are presently being shown, thus no user evaluation has been made. (Task Group 24)
"Predators in Peril" Filmstrip—has been distributed to all schools in the county; too early for user evaluation. (Task Group 25)

Sickle Cell Anemia Survey and Identification. Brochures were developed explaining the disease, and hundreds were requested by school and community groups after the initial distribution was made. Also, an article was published

* IREM published the newsletter for 1 year.

in *Environmental News and Views* concerning the disease, and many requests were made for that issue of the newsletter. (Task Groups 10 and 20)

Communitywide Conference. This involvement process was canceled when it became apparent that the purpose of the conference was being achieved through the CEE Advisory Council.

School Support. Two curriculum development projects relating to environmental preserves were supported by the CEE. The field manuals that were created were distributed to all schools in the county, teachers have taken field trips to the preserve, and other requests for the manuals have been made and honored. (Task Group 29)

The conclusion is that through the involvement of a community's citizenry, awareness and action programs can be developed, implemented, and evaluated to show that environmental education takes place.

Transferability

If an organization's leaders have the ability to facilitate group processes, our model can be utilized in other communities. To assure that people can understand how to become involved, goals, objectives, and specific tasks must be thoroughly articulated. Once this understanding comes about, however, more and more people will want to be able to make a contribution.

Many of the projects were continued even though second year's funding was denied. These projects include the continued publishing of *Environmental News and Views,* the Ecology Centre acting as a teacher resource center, the formation of the Environmental Action Council, and the emergence of an Intergovernmental Environment Council. Schools have continued the community/school workshops, created ecology clubs, nature centers, local curriculum guides, and recycling projects. The county Department of Education is developing curriculum materials for the use of all the schools in the county.

The model and example described here pertain to a county region. A statewide version is being carried out in Colorado, and a model is being designed and implemented for a large city in Los Angeles County. The Environmental Protection Agency serves as a national model.

It can be generalized that none of the models mentioned can survive on volunteer help alone. Many can get started with a large proportion of volunteer help compared to paid staff members. However, it cannot be expected that volunteer help will remain in that status for more than two years. One of two things will happen: (1) participants will become paid staff members and other volunteer help will take their places; (2) workers will resign, and new volunteer help will take their places. The important goal should be education of environmental issues through the participation of people—students and adults alike.

Appendix Description of the 31 Projects with Respect to Target Populations

Target Population: The Community At Large

To educate approximately 1.3 million people about the environment, it becomes necessary to resort to mass media techniques. To that end, the following tasks have been identified:

Task Group 1. Organize and conduct adult school courses through 16 adult schools, 7 community colleges, and 2 university extensions.

Task Group 2. Select presentations by authoritative speakers on environmental topics that are relevant to San Diego County and its development.

Task Group 3. Create television and radio programs by videotaping the speakers from Task 2, which is an effective means of continued environmental education. Also, these programs can be used by service groups in the community.

Task Group 4. It is important to reach as many people as possible through local newspapers. This task is discharged by creating for local newspapers a weekly column, news releases, and bulletins regarding environmental issues.

These four task groups shall work together to explore unifying themes for community consideration resulting in expanded notions of the environment. Examples of such consideration might be the Swedish concept of public access to all property, or the alternative futures concept incorporated in Stanford Research Institute's document, "Alternative Futures and Educational Policy" (Memorandum Report EPRC 6747-6, February, 1970).

Target Population: Community Decision Makers

Decision makers base their decisions on a cost-benefit-efficiency (government) or a cost-benefit-efficiency-profit (business) analysis. In either case economic costs must be considered. It is intended that the following task groups collect and disseminate data reflecting environmental costs that are being incurred if various business or governmental activities interrupt local ecological niches or the environment in general.

Task Group 5. A natural resources inventory was started in San Diego County by the IREM project. Areas that have not been researched in the past have been identified, and these will serve as study areas for students and interested agencies. In this proposal a pilot group of secondary students work with a teacher to complete a natural resources inventory of their community, collate their data, and present their findings to their city council. Other uses for the natural resources inventory begun by IREM are noted for task groups 7, 8, and 9.

Task Group. 6. A community survey needs to be conducted that will determine what the community thinks environmental education *is* and what it *ought* to achieve. The survey will ask questions in four main categories: (1) What is perceived as environmental education? (2) What facts should be known about the condition of the environment? (3) What actions would individuals support to correct unbalanced environmental conditions? (4) What personal values might have to be changed to solve environmental problems? This survey will be made of all four target groups and all geographic areas sampled.

The content vehicle for the questionnaires will revolve around three main categories: (1) preservation of the natural vegetation (land use, open space), (2) human ecology, (3) pollution. This information will provide baseline data for the future assessment of the CEE's effectiveness.

The IREM project is going to make a review of any similar surveys done of the San Diego region in cooperation with the CEE.*

Task Group 7. Workshops to discuss environmental problems will be held for local developers and planners by IREM. The steps to be used during these discussions are: (1) identification problem, (2) define problem quantitatively, (3) collect interpretations of the definitions, (4) list alternative actions for solving the problem.†

Task Group 8. Workshops conducted for government officials. (See Task Group 7 for details.)

Task Group 9. Workshop conducted for leaders of small businesses and corporate executives. (See Task Group 7 for details.)

Task Group 10. Expansion of the CEE's involvement with black and Spanish-speaking community members. Leaders in these community segments have been contacted and encouraged to submit action proposals under the "Human Ecology" component of the CEE program. In response to this, a proposal for a sickle cell anemia information and testing project was submitted. The results of this project will be the testing and education of 7000 blacks. Also, selected environmental education materials will be translated into Spanish.

Task Group 11. Several government agencies (IREM, Water Resources

* CEE survey and IREM review were completed.

† These discussions were never held.

Board, etc.) are conducting studies with regard to the environment. This task is to collect and review the materials and make them available to the other CEE projects.

Target Population: Service Organizations

The tasks groups are formed mainly to provide for community interaction on topics of general concern. CEE's Advisory Council meetings already has had an impact on the community in terms of greater organization interaction and increased mutual awareness. Therefore, the Advisory Council has proposed the following tasks:

Task Group 12. Compile and create a file system for quick access of local and commercial environmental education resources—printed, audiovisual, and human. Through the use of consultants, a clearinghouse will be created and operated. The San Diego County Department of Education is a repository for ERIC and will be available as a resource for this project. Contact has been made with the special environmental education team at Ohio State University. This information will be made available to all citizens and other CEE projects through Task Groups 13 and 14.

Task Group 13. The newsletter from the first year's project will be maintained and its circulation increased to the expanding community resource readership.

Task Group 14. Because so much local information will be collected through the Natural Resource Inventory project, the Community Perception Survey project, the Compilation and Research of Environmental Education Materials project (Task Groups 5, 6, and 24, respectively) and because of Task Group 12, it will be feasible to make available a telephone answering service directing individuals to the wealth of printed, audiovisual, and other information regarding San Diego's environment. One goal of this effort is to make materials available in mimeo or offset form and to disseminate the requested information within 24 hours after a call is received. A computer teletype terminal will be started to aid in retrieving and printing out information.

Task Group 15. Scripps Institution of Oceanography hosted the second public symposium on technical information about the environment. In 1972 the symposium made known the available research regarding pollution. In 1973 the topic was marine life cycles, which is relevant to San Diego because the area has two main outfalls: one discharging treated sewage, the second discharging heated water.

Task Group 16. The continuation of community/school workshops will be held in cooperation with the San Diego City-County Camp Commission at campsites in the Laguna Mountains and will provide an opportunity for parents, teachers, and students to interact and learn more about man's environment.

Task Group 17. Having environmental expositions (fairs) in various shopping centers serves to expose a large number of citizens to the ideas and actions of both student and adult environmentalists. Teachers are encouraged to have students participate by making this an out-of-the-classroom project. Students who do not have a project are urged to evaluate the exposition by critiquing the exhibits for accuracy and perspective. This evaluation then serves as a basis for class discussion.

Task Group 18. The Zoological Society of San Diego has developed an 1800-acre Wild Animal Park. This project is designed to preserve endangered animal species by providing a large natural area for them, hoping that they will reproduce. The educational emphasis of the park is to inform citizens of their place in an ecological system that ought not be put out of balance. It is important that teachers be offered an opportunity to inspect the park firsthand on organized field trips. The Wild Animal Park is one example of many areas students and citizens should see as a part of the workshops they attend.

Task Group 19. A mobile environmental lecture/display will be created to educate students and public alike about San Diego's biotic communities. The display would also make known the many CEE projects with which citizens can become involved. As presently visualized, the project will include leasing a mobile van, selecting various audiovisual aids (some locally produced through Center for Educational Resources (CER)), and provision of a guide/interpreter to respond to questions and give special short talks on a regular schedule. The lecturer will cooperate with the two mobile career guidance vans that presently are operating in the region.

Task Group 20. The Advisory Council plays an important role in project submissions. This task is set up to accomplish that phase of the work.

Task Group 21. Organize and conduct a conference that discusses the state of the art of the recycling enterprises both locally and nationally. This type of program was undertaken this year under the cosponsorship of the Coca-Cola Bottling Company, CER, CEE, and the San Diego Unified School District for 200 teachers. The success of this effort has prompted a desire to repeat the program.

Task Group 22. This task is to document and produce in a suitable media form the CEE's involvement process of environmental education/community/school relations. It is intended that this be done as the program develops.

Target Population: Educators

The tasks directed to the educating segment of San Diego's citizens are designed to (1) provide school systems with a link to the thinking of the community, (2) extend the community environmental education concerns to the various disciplines in the schools through appropriate curriculum development, and (3) utilize existing curricular development units in the area of environmental education.

Task Group 23. Goals and objectives for a K–12 interdisciplinary environmental education curriculum will be based on a survey of local needs. (See Task Group 6). A local school district has done a thorough piece of work in this area based on Paul Brandwein's *Ekistics,* which is serving as an environmental education guideline for California schools.

Task Group 24. Compile, research, and evaluate locally and nationally produced environmental education materials. This will avoid duplication of effort, the material will become part of the Clearinghouse and Dissemination projects (Task Groups 12 and 13, respectively).

Task Group 25. Based on the development of goals and objectives (Task Group 23), where research material is lacking, especially with regard to local needs, development and production of environmental education materials will ensue. The CEE will not develop materials already produced that meet local needs. The resources and experience of the CER section will be utilized in this endeavor.

Task Group 26. Organize and conduct teacher workshops that will lead to the utilization of evaluated curriculum materials on environmental education in the classroom. Whenever possible, arrangements will be made to provide college or university credit for these workshops at teachers' expense.

Task Group 27. To help the CEE keep abreast of national trends, it is important that the staff and other designated citizens attend relevant conferences and symposia. In many instances the CEE personnel will be able to make increasingly significant contributions to these conferences. Also, a major connection has been made between the AAAS Consortium on Environmental Education and the CEE. This link needs to be maintained.

Task Group 28. Students and teachers who become involved in significant activities related to environmental education will be recognized in the President's Environmental Merit Awards Program. Information about projects, special studies, and community action program needs will be made available, and teachers and students will be able to select activities of significance to themselves.

Task Group 29. Some community education efforts in environmental education require supplementary support. One example included in this proposal is the provision for an environmental education specialist on the Scripps Aquarium-Museum staff to develop materials, displays, and student guidesheets emphasizing the environmental education component of their facility to more than 40,000 school children each year. These materials will have broad usage for other students through their incorporation by the K–12 materials development task. (Task Groups 23 and 25)

Task Group 30. School administrators have begun to be questioned by parents about the content of the environmental education programs. Administrative personnel will need assistance in explaining the derivation of environmental education efforts (values, content, methodology, etc.); for this, specific workshops will be planned.

Task Group 31. The CEE has received many requests from school district curriculum coordinators as well as teachers asking for help in setting up environmental education experiences. Thus resource teachers who will be able to respond to these requests will be placed on call. They will have the full complement of resources described in the other tasks of this project to carry out this assignment. From the evaluated review of these requests, information can be derived to suggest specific programs that will need to be developed in the future.

Regional Environmental Management Data

BRUNO LORAN

Abstract

Regional environmental management data apply to broad environmental subjects, such as air, water, thermal, and radioactive pollution, solid waste management, pollutant monitoring, and energy generation and use. They apply also to regional aspects, such as land use, environmental impact studies, regional zoning according to ecological principles, resources allocation, and state-of-the-environment indices. Many of the needed data already exist or have been obtained for similar situations, and the results can be transferred to regional settings. The problem is to become familiar with the data sources that may apply and to retrieve the desired information from these sources. In this chapter, the sources that are applicable to environmental management are reviewed and discussed.

Within the last five years, many of the data sources mentioned have become computerized; that is, the information has been stored on computer tapes, which may be searched using logic systems that permit pinpointing specific subjects or subject areas. State or federal information-retrieval centers conduct computer searches for private users; the most important centers are listed.

At a recent EPA-sponsored National Environmental Information Symposium, William D. Ruckelshaus remarked:

We may indeed, as one scientist has suggested, already have a substantial part of the scientific information we need to ensure the protection and preservation of our common

Mr. Loran is with the Western Research Application Center of the University of Southern California at Los Angeles.

environment. Yet we don't have timely access to it because retrieval systems are uncoordinated or nonexistent. There is as great a need to organize and manage information as there is to make new discoveries.

This chapter presents the state of the art on current environmental management information. The access to this information is fairly complex, and this material may serve as a guide to its retrieval.

Sources Overview

Regional environmental management involves the application of broad environmental principles and disciplines to specific local situations. The general concepts utilized are the collection of environmental data (especially concerning air and water), the setting up of monitoring stations to follow the variation of data over periods of time, the analysis of the data obtained, the formulation of pollution control policies directed at air, water, thermal, and radioactive pollution, the inventory of available resources (including intangible ones, such as recreational areas and "open spaces"), the generation and use of energy, and the management of solid wastes. Aspects more specifically regional are land use, preservation of the natural habitat in areas of ecological interest (such as deserts or swamps), environmental impact studies, regional zoning according to sound environmental principles, resources allocation, state-of-the-environment indices, and the study of how federal environmental standards and criteria apply to local conditions.

Competent decisions cannot be made about any of the subjects mentioned without knowledge of engineering techniques, social factors, and legal considerations. How can this knowledge be acquired? The results of technical and social studies concerning the environment, both at a national and a regional level, have appeared in many books, reports, and journals. They have been collected in abstract form by government agencies, such as the National Technical Information Service, or private groups, such as the Engineering Societies Library and the American Chemical Society, which assemble and publish "data sources" respectively entitled: *Government Reports Announcements, Engineering Index,* and *Chemical Abstracts.* By consulting these sources, one can obtain access to information on the subjects desired. No study or research project should be initiated without prior consultation of the pertinent sources. It is possible that exactly the same study has already been carried out; but even if this is not the case, studies may have been performed under similar conditions. The GRA data bank, for example (see below), systematically abstracts all Environmental Impact Statements that have been written (close to 5000 at this writing); some of these are bound to parallel local conditions being investigated.

Sources For Abstracts

Environmental subjects are covered by the following data sources, which are stored in most university libraries.

Government Reports Announcements (GRA). GRA announces reports of research and development sponsored by the federal government. It includes all unclassified reports and translations released to the public through the National Technical Information Service (NTIS) by government departments and agencies; it gives, however, only partial coverage to AEC and NASA material. The index to GRA is published separately (under the name of *Government Reports Index*). The GRA data bank is a good starting point for an environmental search. Other sources may contain more information in specific fields and are also not limited to government-sponsored research.

Air Pollution Abstracts (sponsored by EPA). International coverage of air pollution-related subjects.

Selected Water Resources Abstracts (sponsored by the Office of Water Resources Research). International coverage of subjects related to water pollution and water management. Information on ongoing research projects in the same fields is contained in the *Water Resources Research Catalog*, also sponsored by the Office of Water Resources Research.

Environment Index. Provides quick retrieval of pertinent environmental literature through a list of titles, issued yearly. Abstracts and full copies may be ordered from the publisher (Environment Information Center, Inc., 124 East 39th St., New York, New York, 10016).

Engineering Index. Worldwide coverage of all engineering and technology aspects, including environmental applications.

Chemical Abstracts. International coverage of the chemical field. Special sections: "Air Pollution," "Water," "Sewage and Wastes."

Biological Abstracts. Worldwide coverage of the biological and biochemical literature. Offers a very good review of the biological effects of pollutants.

NASA Material. Published in two sections, *Scientific and Technical Aerospace Reports* (STAR), containing abstracts of NASA-sponsored studies, and *International Aerospace Abstracts* (IAA), containing abstracts from the open literature. In recent years, NASA has dedicated a considerable amount of research to the environment, especially in the fields of pollutant analysis and remote sensing.

Nuclear Science Abstracts (sponsored by AEC). Complete coverage of radioactive pollution and radioactive waste subjects.

Oceanic Index. Worldwide coverage of ocean-related subjects, including ocean pollution and offshore development.

Index Medicus (sponsored by the U.S. Department of Health, Education, and Welfare). Worldwide coverage of the medical literature, including health-related aspects of pollution.

Solid Waste Literature Abstracts (sponsored by EPA). Subjects related to solid waste management and solid waste research. Compiled by the Sanitary Engineering Research Laboratory of the University of California at Berkeley (limited circulation, may not be available in all libraries).

Pollution Abstracts. Source of limited value because both technical and nontechnical (newspaper and magazine articles) material is abstracted.

All the article and book sources previously mentioned provide abstracts (titles for *Index Medicus*). The full reports may be available from the issuing agency or from local libraries. Usually information on document availability is contained in the first pages of the publications. A good source of information on journal holdings by libraries is *Access*, published by the American Chemical Society.

Sometimes a full text on environmental subjects is desired, rather than a report or paper on specific aspects. In this case, the *Library of Congress Catalog* (Books: Subjects) may be consulted; all books published everywhere in the world are listed, including publications of the Government Printing Office. Much more restricted coverage is provided by the subject index of *Books in Print*, published by the R. R. Bowker Company in New York. English-language books printed by private publishers are listed.

Environmental Law Sources

Many of the sources listed cover legal aspects of environmental subjects. Specific legislative and regulatory information may be obtained directly from the official publications of the legislative bodies (e.g., *The Federal Register*, which is stored in all university law libraries and in many public libraries). A private organization, the Bureau of National Affairs (Washington, D.C.), systematically collects and republishes in *The Environmental Reporter* federal and state environmental laws and regulations; many libraries subscribe to this publication.

Mainly since 1968, the information contained in the data sources has been put on magnetic tape, either as full abstracts (for the GRA data bank only) or as citations (i.e., title, reference, and a series of "descriptors" or "key words" describing the subject). Computer searching is more efficient than manual searching and is also much faster. Computerized retrieval of the information occurs through the key words, which are assigned labels and treated as mathematical symbols, using Boolean algebra. The retrieval procedure, or "strategy," is very simple. Three operators are used: the + or OR logic, which permits retrieval of all citations containing the key word; the · or AND logic, which pairs key words together, permitting only citations containing both key words to be retrieved; and the − or NOT logic, which inhibits the retrieval or undesired key words. These forms are illustrated in Figure 1.

Following is an example of a simple computer strategy; the information desired is "regional aspects of air and water pollution":

A	Air pollution
B	Water pollution
C	County
D	State
E	Region
F	Regional environmental management
G	Federal

Strategy: $(A + B) \cdot (C + D + E) + F - G$

Of the data sources mentioned, GRA, *Air Pollution Abstracts, Selected Water Resources Abstracts, Engineering Index, Chemical Abstracts, Biological Abstracts*, the NASA Materials, *Nuclear Science Abstracts*, and *Index Medicus* are available on tape. The *Water Resources Research Catalog* is included in a larger-scope data bank produced by the Smithsonian Institution and covering information on ongoing research projects in United States.

If a very large volume of environmental searches is expected and a computer is available, the tapes may be purchased or leased (average price: $3000 per year). Computer searches may also be obtained from information retrieval and dissemination centers, operating with state or federal support, as follows.

Air Pollution Abstracts searches are available, free of charge, from the Air Pollution Technical Information Center (APTIC), Research Triangle Park,

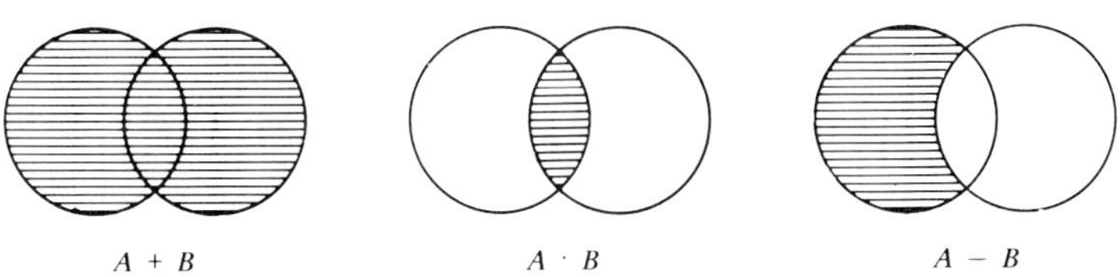

Figure 1. Graphical representation of computer logic for information retrieval.

North Carolina. *Index Medicus* computer searches are available, also without charge, from most medical university libraries. GRA searches are run for a $50 charge by the National Technical Information Service (NTIS), Springfield, Virginia (output limited to 100 abstracts). *Selected Water Resources* may be searched for a $35 fee through the University of Wisconsin,-Cornell University, Ithaca, New York; or North Carolina State University, Raleigh. Searches for *ongoing research projects* in specific fields are provided by the Smithsonian Institution, Washington, D.C., for approximately $50.

GRA, *Engineering Index, Chemical Abstracts*, and NASA material searches are run for an average $150 charge per data bank (no output limit) by six regional information-retrieval and dissemination centers operating with NASA support: Aerospace Research Application Center (ARAC), University of Indiana, Bloomington; Knowledge Availability Systems Center (KASC), University of Pittsburgh; New England Research Application Center (NERAC), University of Connecticut, Storrs; North Carolina Science and Technology Research Center (NCSTRC), Research Triangle Park; Technology Application Center (TAC), University of New Mexico, Albuquerque; Western Research Application Center (WESRAC), University of Southern California, Los Angeles. These centers also provide manual searches. A comprehensive computerized information-retrieval service is provided, at somewhat higher charges, by the Computer Center Information Services of the University of Georgia (Athens).

Summary

It is seen that information on environmental management must now be obtained from fragmented sources, according to the subject and the specific field of interest. It is hoped that this chapter will be helpful to those who need to retrieve environmental data. EPA is currently considering a centralized Environmental Data Bank that could be tapped by "one line," using their own terminals. Such a system would satisfy the need expressed by Mr. Ruckelshaus. However, the environmental field is so complex, and covers so many different disciplines, that the organization of such a centralized system is believed to be a very difficult task.

Index